This item must be returned or renewed by the last date shown below. The loan period may be shortened if it is reserved by another reader. A fine may be due if it is not returned on time.

DATE OF RETURN

11 JAN 1999

02 MAY 2000

10 SEP 2004

09 JUN 2006

25 JAN 2008

UNIVERSITY COLLEGE LONDON

Gower Street London WC1E 6BT

LF4D

Vibrations of Elastic Plates

Springer
New York
Berlin
Heidelberg
Barcelona
Budapest
Hong Kong
London
Milan
Paris
Santa Clara
Singapore
Tokyo

Yi-Yuan Yu

Vibrations of Elastic Plates

Linear and Nonlinear Dynamical Modeling of Sandwiches, Laminated Composites, and Piezoelectric Layers

With 27 Illustrations

Springer

Yi-Yuan Yu
Professor Emeritus
New Jersey Institute of Technology
Newark, NJ 07021
USA

Library of Congress Cataloging-in-Publication Data
Yu, Yi-Yuan, 1923–
    Vibrations of elastic plates : linear and nonlinear dynamical
    modeling of sandwiches, laminated composites, and piezoelectric
    layers / Yi-Yuan Yu.
        p. cm.
    Includes bibliographical references and index.
    ISBN 0-387-94514-8 (alk. paper)
    1. Plates (Engineering) — Vibration — Mathematical models.
    2. Sandwich construction — Vibration — Mathematical models.
    3. Composite construction — Vibration — Mathematical models.
    I. Title.
    TA492.P7Y6   1995
    824.1′7765 — dc20   95-8375

Printed on acid-free paper.

Production supervised by Karen Phillips and managed by Publishing Network.
Manufacturing supervised by Jeffrey Taub.
Typeset by Bytheway Typesetting Services, Inc., Norwich, NY.
Printed and bound by Braun-Brumfield, Ann Arbor, MI.
Printed in the United States of America.

9 8 7 6 5 4 3 2 1

ISBN 0-387-94514-8 Springer-Verlag New York Berlin Heidelberg   SPIN 10500078

To Eileen,
Yolanda, Lisa, and John

Preface

This book is based on my experiences as a teacher and as a researcher for more than four decades. When I started teaching in the early 1950s, I became interested in the vibrations of plates and shells. Soon after I joined the Polytechnic Institute of Brooklyn as a professor, I began working busily on my research in vibrations of sandwich and layered plates and shells, and then teaching a graduate course on the same subject. Although I tried to put together my lecture notes into a book, I never finished it. Many years later, I came to the New Jersey Institute of Technology as the dean of engineering. When I went back to teaching and looked for some research areas to work on, I came upon laminated composites and piezoelectric layers, which appeared to be natural extensions of sandwiches. Working on these for the last several years has brought me a great deal of joy, since I still am able to find my work relevant. At least I can claim that I still am pursuing life-long learning as it is advocated by educators all over the country. This book is based on the research results I accumulated during these two periods of my work, the first on vibrations and dynamical modeling of sandwiches, and the second on laminated composites and piezoelectric layers.

Beams, plates, and shells are thin structures that have a dimension much smaller in the thickness direction than in the other directions in the plane or surface of the structure. In the analysis of such elastic structures, there is a choice between the use of the three-dimensional elasticity theory and that of one- and two-dimensional beam, plate, and shell theories. The analysis usually is simpler by applying the latter theories. However, the elasticity theory still serves two important purposes. First, it can be used directly to solve problems of beams, plates, and shells when the problems are simple enough, and the solutions obtained often provide mathematical and physical insight that cannot be attained through other means. Second, the three-dimensional elasticity theory also provides a foundation for the development of one- and two-dimensional beam, plate, and shell theories. Thus, elasticity theory plays an essential part in the study of elastic thin structures, as will be elaborated in this book by treating linear and nonlinear vibrations and dynamical modeling.

The book deals with dynamics and vibration of elastic thin structures, particularly plates. Throughout the book I have made an effort to develop dynamical modeling through an extensive and systematic use of variational equations of motion in both linear and nonlinear elasticity. According to the standard treatise on elasticity by A.E.H. Love, it was G. Kirchhoff who first derived the ordinary variational equation of motion in linear elasticity from Hamilton's principle. Many years have elapsed since then, as I introduced a generalized variational equation of motion in nonlinear elasticity in 1964, proposed the use of pseudo-variational equations of motion in 1991, and further included the piezoelectric effect in 1994 and 1995. These linear and nonlinear variational equations of motion are indeed very powerful tools for dynamical modeling, as will be demonstrated throughout this book.

Thus, Chapter 1 provides the fundamental elements of nonlinear elasticity theory for large deformations, including linear elasticity for small deformations as a special case. These are followed by the development of the ordinary, generalized, and pseudo-variational equations of motion. From this chapter onward, the book covers essentially two parts, the first part on linear and the second part on nonlinear vibrations and dynamical modeling of elastic plates.

The first part of the book includes Chapters 2 through 6. In Chapter 2, linear vibrations of plates are treated through the use of linear elasticity theory itself. Linear dynamical modeling then is covered in the next three chapters for homogeneous, sandwich, and laminated composite plates. Based on the use of the variational equation of motion in linear elasticity, both classical and refined plate equations of motion are developed for all of these types of plates. The ranges of usefulness of these equations in vibration analysis are examined and determined by investigating vibrations of an infinite plate. Applications to finite-sized plates are discussed in Chapter 6.

The second part of the book consists of the remaining chapters, which are devoted to nonlinear vibrations and dynamical modeling for elastic plates with large deflections. Chapter 7 deals with homogeneous beams, plates, and shallow shells, and Chapter 8 with sandwich and laminated composite plates. A modern treatment of nonlinear vibrations would not be complete without considering chaos, which is introduced in Chapter 9 through the use of the well-known Duffing equation. The emphasis is on the effect of small damping, and some numerical results are shown for the first time in this book. In the final chapter of the book, nonlinear dynamical modeling for large deflections of piezoelectric plates is treated, with some of the concepts introduced only recently. This will find applications to distributed vibration control of flexible thin structures, and represents a discipline that has become increasingly active and important again as new structural and material systems are adopted in modern engineering practice. In any case, the use of the variational equations of motion appears to be a logical way of extending the treatment in this book to the numerical analysis of vibrations of elastic plates.

I would like to give my thanks to Thomas von Foerster of Springer-Verlag for suggesting the final title of the book, and to Bernard Koplik of the New Jersey Institute of Technology, my long-time friend and former student, for going over my manuscript before publication.

Yi-Yuan Yu
December 1995

Contents

1

Nonlinear Elasticity Theory

For many years, a standard treatise on elasticity has been the book by Love (1927). Another standard text on linear elasticity has been prepared by Timoshenko and Goodier (1970), which first appeared in 1934. Among other books on nonlinear elasticity, we mention those by Novozhilov (1948) and Fung (1965). Both of these cover the classical nonlinear case in detail. The book by Novozhilov also deals with a simplified nonlinear case of small strains and large rotations that has found important applications to large deflections of thin structures by many authors. The treatise by Fung further deals with foundations of solid mechanics in general. His book includes a very well-prepared bibliography, covering the literature in solid mechanics before 1965.

In this first chapter, the elements of nonlinear elasticity theory for large deformations, including linear elasticity for small deformations as a special case, are presented. Strains and stresses are discussed in Sections 1.1 and 1.2, respectively, in which the Green nonlinear strain tensor and Kirchoff stress tensor are introduced. The important concepts of the strain energy function and the principle of virtual work are discussed in Section 1.3. The principle of virtual work is next extended to the dynamic case in Section 1.4, in which Hamilton's principle and the associated variational equation of motion in nonlinear elasticity are formulated (Yu 1964). The latter is sometimes referred to in this book as the ordinary variational equation of motion. The discussion is then further extended to cover the pseudo-variational equations of motion in Section 1.5 (Yu 1991), and the generalized Hamilton's principle and associated generalized variational equation of motion in Section 1.6 (Yu 1964). Elastic stress–strain relations for both small and large deformations are discussed in the last section in this chapter. The piezoelectric effect in a three-dimensional nonlinear theory will be treated in Chapter 10 (Yu 1955a,b).

1.1   Strains

Under the action of external forces, a body undergoes a change in position. If the body is elastic, it also deforms and strains are developed. In the analysis of strain, we first introduce Green's strain tensor, which is associated with the classical case in nonlinear elasticity. This is next reduced to a simplified nonlinear case and finally to the linear case.

1.1.1   Green's Strain Tensor

We begin with a description of the change in distance between two points in the body. This can be expressed in terms of the coordinates before or after the deformation. Here, the coordinates before deformation are used. Consider two points A and B in a body in its undeformed state, as shown in Figure 1.1.1. Their positions are described by the vectors $OA = \mathbf{r}$ and $OB = \mathbf{r} + d\mathbf{r}$, respectively. Let A' and B' be the new positions of A and B after deformation. The displacement vectors from the original to the new positions are then $AA' = \mathbf{u}$ and $BB' = \mathbf{u} + d\mathbf{u}$.

Let $\mathbf{i}_x$, $\mathbf{i}_y$, and $\mathbf{i}_z$ be the unit vectors in the original directions of the rectangular coordinates x, y, z, and let u_x, u_y, and u_z be the displacement components in these directions. Thus, the position and displacement vectors of the point A are defined by, respectively,

$$\mathbf{r} = \mathbf{i}_x x + \mathbf{i}_y y + \mathbf{i}_z z$$
$$\mathbf{u} = \mathbf{i}_x u_x + \mathbf{i}_y u_y + \mathbf{i}_z u_z,$$

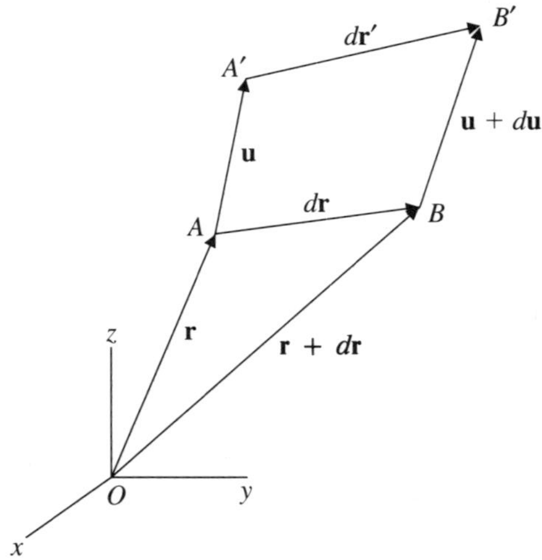

Fig. 1.1.1. Deformation in a solid.

and the position vector of the point A' is defined by

$$\mathbf{r}' = \mathbf{r} + \mathbf{u} = \mathbf{i}_x(x + u_x) + \mathbf{i}_y(y + u_y) + \mathbf{i}_z(z + u_z).$$

It follows that the line elements before and after deformation are, respectively,

$$AB = d\mathbf{r} = \mathbf{i}_x dx + \mathbf{i}_y dy + \mathbf{i}_z dz \tag{1.1.1}$$

$$A'B' = d\mathbf{r}' = d\mathbf{r} + d\mathbf{u}$$

$$= d\mathbf{r} + \left(\frac{\partial \mathbf{u}}{\partial x}\right) dx + \left(\frac{\partial \mathbf{u}}{\partial y}\right) dy + \left(\frac{\partial \mathbf{u}}{\partial z}\right) dz$$

$$= \left[\mathbf{i}_x\left(1 + \frac{\partial u_x}{\partial x}\right) + \mathbf{i}_y\left(\frac{\partial u_y}{\partial x}\right) + \mathbf{i}_z\left(\frac{\partial u_z}{\partial x}\right)\right] dx$$

$$+ \left[\mathbf{i}_x\left(\frac{\partial u_x}{\partial y}\right) + \mathbf{i}_y\left(1 + \frac{\partial u_y}{\partial y}\right) + \mathbf{i}_z\left(\frac{\partial u_z}{\partial y}\right)\right] dy \tag{1.1.2}$$

$$+ \left[\mathbf{i}_x\left(\frac{\partial u_x}{\partial z}\right) + \mathbf{i}_y\left(\frac{\partial u_y}{\partial z}\right) + \mathbf{i}_z\left(1 + \frac{\partial u_z}{\partial z}\right)\right] dz$$

$$= \mathbf{g}_x dx + \mathbf{g}_y dy + \mathbf{g}_z dz,$$

where $\mathbf{g}_x$, $\mathbf{g}_y$, and $\mathbf{g}_z$ are the base vectors of the deformed body defined by

$$\mathbf{g}_x = \frac{\partial \mathbf{r}'}{\partial x} = \mathbf{i}_x + \frac{\partial \mathbf{u}}{\partial x} = \mathbf{i}_x\left(1 + \frac{\partial u_x}{\partial x}\right) + \mathbf{i}_y\left(\frac{\partial u_y}{\partial x}\right) + \mathbf{i}_z\left(\frac{\partial u_z}{\partial x}\right)$$

$$\mathbf{g}_y = \frac{\partial \mathbf{r}'}{\partial y} = \mathbf{i}_y + \frac{\partial \mathbf{u}}{\partial y} = \mathbf{i}_x\left(\frac{\partial u_x}{\partial y}\right) + \mathbf{i}_y\left(1 + \frac{\partial u_y}{\partial y}\right) + \mathbf{i}_z\left(\frac{\partial u_z}{\partial y}\right) \tag{1.1.3}$$

$$\mathbf{g}_z = \frac{\partial \mathbf{r}'}{\partial z} = \mathbf{i}_z + \frac{\partial \mathbf{u}}{\partial z} = \mathbf{i}_x\left(\frac{\partial u_x}{\partial z}\right) + \mathbf{i}_y\left(\frac{\partial u_y}{\partial z}\right) + \mathbf{i}_z\left(1 + \frac{\partial u_z}{\partial z}\right).$$

These are the new vectors into which the unit vectors $\mathbf{i}_x$, $\mathbf{i}_y$, and $\mathbf{i}_z$ are transformed after the deformation.

We next take the difference between the squares of the line elements $d\mathbf{r}'$ and $d\mathbf{r}$:

$$d\mathbf{r}' \cdot d\mathbf{r}' - d\mathbf{r} \cdot d\mathbf{r} = (\mathbf{g}_x^2 - 1)(dx)^2 + (\mathbf{g}_y^2 - 1)(dy)^2 + (\mathbf{g}_z^2 - 1)(dz)^2$$

$$+ 2\mathbf{g}_x \cdot \mathbf{g}_y\, dx\, dy + 2\mathbf{g}_y \cdot \mathbf{g}_z\, dy\, dz + 2\mathbf{g}_z \cdot \mathbf{g}_x\, dz\, dx. \tag{1.1.4}$$

The coefficients in Eq. (1.1.4) are associated with an arbitrary line element, which has the components dx, dy, and dz before deformation. They are used to define the nonlinear strain components as follows:

$$\epsilon_{xx} = \tfrac{1}{2}(\mathbf{g}_x \cdot \mathbf{g}_x - 1) = \frac{\partial u_x}{\partial x} + \frac{1}{2}\left[\left(\frac{\partial u_x}{\partial x}\right)^2 + \left(\frac{\partial u_y}{\partial x}\right)^2 + \left(\frac{\partial u_z}{\partial x}\right)^2\right]$$

$$\epsilon_{yy} = \tfrac{1}{2}(\mathbf{g}_y \cdot \mathbf{g}_y - 1) = \frac{\partial u_y}{\partial y} + \frac{1}{2}\left[\left(\frac{\partial u_x}{\partial y}\right)^2 + \left(\frac{\partial u_y}{\partial y}\right)^2 + \left(\frac{\partial u_z}{\partial y}\right)^2\right]$$

$$\epsilon_{zz} = \tfrac{1}{2}(\mathbf{g}_z \cdot \mathbf{g}_z - 1) = \frac{\partial u_z}{\partial z} + \frac{1}{2}\left[\left(\frac{\partial u_x}{\partial z}\right)^2 + \left(\frac{\partial u_y}{\partial z}\right)^2 + \left(\frac{\partial u_z}{\partial z}\right)^2\right]$$

$$\epsilon_{xy} = \tfrac{1}{2}\mathbf{g}_x \cdot \mathbf{g}_y = \frac{1}{2}\left[\frac{\partial u_y}{\partial x} + \frac{\partial u_x}{\partial y} + \left(\frac{\partial u_x}{\partial x}\right)\left(\frac{\partial u_x}{\partial y}\right)\right.$$
$$\left. + \left(\frac{\partial u_y}{\partial x}\right)\left(\frac{\partial u_y}{\partial y}\right) + \left(\frac{\partial u_z}{\partial x}\right)\left(\frac{\partial u_z}{\partial y}\right)\right] \tag{1.1.5}$$

$$\epsilon_{yz} = \tfrac{1}{2}\mathbf{g}_y \cdot \mathbf{g}_z = \frac{1}{2}\left[\frac{\partial u_z}{\partial y} + \frac{\partial u_y}{\partial z} + \left(\frac{\partial u_x}{\partial y}\right)\left(\frac{\partial u_x}{\partial z}\right)\right.$$
$$\left. + \left(\frac{\partial u_y}{\partial y}\right)\left(\frac{\partial u_y}{\partial z}\right) + \left(\frac{\partial u_z}{\partial y}\right)\left(\frac{\partial u_z}{\partial z}\right)\right]$$

$$\epsilon_{zx} = \tfrac{1}{2}\mathbf{g}_z \cdot \mathbf{g}_x = \frac{1}{2}\left[\frac{\partial u_x}{\partial z} + \frac{\partial u_z}{\partial x} + \left(\frac{\partial u_x}{\partial z}\right)\left(\frac{\partial u_x}{\partial x}\right)\right.$$
$$\left. + \left(\frac{\partial u_y}{\partial z}\right)\left(\frac{\partial u_y}{\partial x}\right) + \left(\frac{\partial u_z}{\partial z}\right)\left(\frac{\partial u_z}{\partial x}\right)\right].$$

These can be shown to be the components of a symmetric tensor, often called *Green's strain tensor*. The components are directly related to the deformations. Thus, consider first a line element lying originally in the x-direction. The extensional strain ϵ_x in this direction is readily shown to be related to the normal strain component ϵ_{xx} by

$$\epsilon_x = \sqrt{1 + 2\epsilon_{xx}} - 1.$$

Consider next two line elements that originally form a right angle with each other, say, in the x- and y-directions. The new angle θ between the line elements after deformation is a measure of the shearing strain component ϵ_{xy}, given by the expression

$$\cos\theta = \frac{2\epsilon_{xy}}{\sqrt{(1 + 2\epsilon_{xx})(1 + 2\epsilon_{yy})}}.$$

The nonlinear strain components $\epsilon_{xx}, \ldots, \epsilon_{xy}, \ldots$ are therefore direct measures of the extensional and shearing strains. Specifically, when these strain components are 0, the corresponding strains also vanish.

1.1.2 Linear Strains and Rotations

Let us introduce the usual linear strains

$$e_{xx} = \frac{\partial u_x}{\partial x}$$
$$e_{yy} = \frac{\partial u_y}{\partial y}$$
$$e_{zz} = \frac{\partial u_z}{\partial z}$$

$$e_{xy} = \frac{1}{2}\left(\frac{\partial u_y}{\partial x} + \frac{\partial u_x}{\partial y}\right)$$

$$e_{yz} = \frac{1}{2}\left(\frac{\partial u_z}{\partial y} + \frac{\partial u_y}{\partial z}\right)$$

$$e_{zx} = \frac{1}{2}\left(\frac{\partial u_x}{\partial z} + \frac{\partial u_z}{\partial x}\right)$$

$$(1.1.6)$$

and the rotations

$$\omega_{xy} = \frac{1}{2}\left(\frac{\partial u_y}{\partial x} - \frac{\partial u_x}{\partial y}\right)$$

$$\omega_{yz} = \frac{1}{2}\left(\frac{\partial u_z}{\partial y} - \frac{\partial u_y}{\partial z}\right)$$

$$\omega_{zx} = \frac{1}{2}\left(\frac{\partial u_x}{\partial z} - \frac{\partial u_z}{\partial x}\right)$$

$$(1.1.7)$$

While the linear strains are also components of a symmetric tensor, the rotations are components of an anti-symmetric tensor.

In the engineering literature, the tensorial linear strains are often replaced by engineering strains, defined as follows:

$$e_x = \frac{\partial u_x}{\partial x}, \cdots$$

$$\gamma_{xy} = \frac{\partial u_y}{\partial x} + \frac{\partial u_x}{\partial y}, \cdots.$$

$$(1.1.8)$$

The engineering extensional strains are thus the same, but the engineering shearing strains are twice as large as the corresponding components of the linear strain tensor. The rotations are sometimes expressed as components of a vector, related to the above components of a rotation tensor as follows:

$$\omega_x = \omega_{yz}, \cdots.$$

$$(1.1.9)$$

These can be interpreted as the mean rotations of a volume element about the x, y, and z axes, respectively.

1.1.3 Classical Nonlinear Case

Equations (1.1.5) represent the classical nonlinear case that may be characterized by *large finite deformations*. In terms of linear strains and rotations, the nonlinear strains in these equations take the following form:

$$\epsilon_{xx} = e_{xx} + \tfrac{1}{2}[e_{xx}^2 + (e_{xy} + \omega_{xy})^2 + (e_{zx} - \omega_{zx})^2]$$

$$\epsilon_{yy} = e_{yy} + \tfrac{1}{2}[e_{yy}^2 + (e_{yz} + \omega_{yz})^2 + (e_{xy} - \omega_{xy})^2]$$

$$\epsilon_{zz} = e_{zz} + \tfrac{1}{2}[e_{zz}^2 + (e_{zx} + \omega_{zx})^2 + (e_{yz} - \omega_{yz})^2]$$

$$(1.1.10)$$

$$\epsilon_{xy} = e_{xy} + \tfrac{1}{2}[e_{xx}(e_{xy} - \omega_{xy}) + e_{yy}(e_{xy} + \omega_{xy}) + (e_{zx} - \omega_{zx})(e_{yz} + \omega_{yz})]$$

$$\epsilon_{yz} = e_{yz} + \tfrac{1}{2}[e_{yy}(e_{yz} - \omega_{yz}) + e_{zz}(e_{yz} + \omega_{yz}) + (e_{xy} - \omega_{xy})(e_{zx} + \omega_{zx})]$$

$$\epsilon_{zx} = e_{zx} + \tfrac{1}{2}[e_{zz}(e_{zx} - \omega_{zx}) + e_{xx}(e_{zx} + \omega_{zx}) + (e_{yz} - \omega_{yz})(e_{xy} + \omega_{xy})].$$

These describe the general situation in which the magnitudes of linear strains and rotations can be arbitrary.

1.1.4 Simplified Nonlinear Case

Thin structures such as beams, plates, and shells are usually so flexible that large rotations can develop even when strains are small. If the linear strains are assumed to be much smaller than the rotations, Eqs. (1.1.10) reduce to the following simplified form (Novozhilov 1948):

$$\epsilon_{xx} = e_{xx} + \tfrac{1}{2}(\omega_{xy}^2 + \omega_{zx}^2)$$

$$\epsilon_{yy} = e_{yy} + \tfrac{1}{2}(\omega_{yz}^2 + \omega_{xy}^2)$$

$$\epsilon_{zz} = e_{zz} + \tfrac{1}{2}(\omega_{zx}^2 + \omega_{yz}^2)$$

$$\epsilon_{xy} = e_{xy} - \tfrac{1}{2}\omega_{zx}\omega_{yz}$$

$$\epsilon_{yz} = e_{yz} - \tfrac{1}{2}\omega_{xy}\omega_{zx}$$

$$\epsilon_{zx} = e_{zx} - \tfrac{1}{2}\omega_{yz}\omega_{xy}.$$

$$(1.1.11)$$

These represent the simplified nonlinear case of small strains and large rotations, which may be characterized by *small finite deformations*. The deformations are still nonlinear, but they must be small so that the range of deformation remains elastic and no permanent set or failure occurs.

1.1.5 Linear Case

When the nonlinear terms become negligible in Eqs. (1.1.5), (1.1.10), and (1.1.11), the nonlinear strains reduce to linear strains, and all rotations disappear. This is, of course, the linear case, which is characterized by the familiar *infinitesimal deformations*.

1.2 Stresses

In the analysis of stress, we introduce first Kirchhoff's stress tensor, which is associated with the classical case of nonlinear elasticity, as in the analysis of strain. This is reduced next to the simplified nonlinear cases and finally to the linear case. Both the equilibrium equations and traction boundary conditions are treated in each of these cases.

1.2.1 Kirchhoff's Stress Tensor

The stress at an interior point in a body depends not only on the surface element on which the stress is acting, but also on the direction and sense of that surface element. This is especially important for large finite deformations because the surface element and its direction can be chosen in the original undeformed state, in the new deformed state, or in any combination of the two. As pointed out by Marguerre (1962), calculations are particularly simple by choosing an interior volume element $dx\,dy\,dz$ from the undeformed state of the body under consideration, referring the forces acting on the faces of the element to their undeformed areas, but resolving the stress vectors into components in the directions of the deformed base vectors $\mathbf{g}_x$, $\mathbf{g}_y$, $\mathbf{g}_z$. Thus, we let $\mathbf{s}_x$, $\mathbf{s}_y$, $\mathbf{s}_z$ be the stress vectors acting on faces that have normals in the x-, y-, and z-directions, respectively, and write the forces on the faces as $\mathbf{s}_x\,dy\,dz$, $\mathbf{s}_y\,dz\,dx$, and $\mathbf{s}_z\,dx\,dy$. When the stress vectors are resolved, we find

$$
\begin{aligned}
\mathbf{s}_x &= \mathbf{g}_x \sigma_{xx} + \mathbf{g}_y \sigma_{xy} + \mathbf{g}_z \sigma_{xz} \\
\mathbf{s}_y &= \mathbf{g}_x \sigma_{yx} + \mathbf{g}_y \sigma_{yy} + \mathbf{g}_z \sigma_{yz} \\
\mathbf{s}_z &= \mathbf{g}_x \sigma_{zx} + \mathbf{g}_y \sigma_{zy} + \mathbf{g}_z \sigma_{zz}.
\end{aligned}
\tag{1.2.1}
$$

The stresses σ_{xx}, σ_{yy}, σ_{zz}, σ_{xy}, σ_{yz}, and σ_{zx} are then the components of a tensor, often called *Kirchhoff's stress tensor*. This tensor is symmetric, since equilibrium of the moments acting on the volume element $dx\,dy\,dz$ can be shown to yield the symmetric relations $\sigma_{xy} = \sigma_{yx}$, $\sigma_{yz} = \sigma_{zy}$, and $\sigma_{zx} = \sigma_{xz}$.

1.2.2 Classical Nonlinear Case

An equilibrium equation for large finite deformations in nonlinear elasticity can be formulated in vector form in a manner similar to that for infinitesimal deformations in linear elasticity. As shown in Figure 1.2.1, the forces on the two opposite faces $x = \text{constant}$ of the volume element are $\mathbf{s}_{-x}\,dy\,dz$ and $[\mathbf{s}_x + (\partial \mathbf{s}_x / \partial x)\,dx]\,dy\,dz$, respectively. Since $\mathbf{s}_{-x} = \mathbf{s}_x$, the net force is equal to $(\partial \mathbf{s}_x / \partial x)\,dx\,dy\,dz$. Similar terms can be written for forces acting on the other two pairs of faces of the volume element. Equilibrium then requires that

$$
\frac{\partial \mathbf{s}_x}{\partial x} + \frac{\partial \mathbf{s}_y}{\partial y} + \frac{\partial \mathbf{s}_z}{\partial z} + \mathbf{f} = 0,
\tag{1.2.2}
$$

where

$$
\mathbf{f} = \mathbf{i}_x f_x + \mathbf{i}_y f_y + \mathbf{i}_z f_z
$$

is the body force vector per unit volume of the undeformed body.

Substituting Eqs. (1.2.1) in (1.2.2) and decomposing the result into components in the $\mathbf{i}_x$-, $\mathbf{i}_y$-, and $\mathbf{i}_z$-directions, we find

$$
\frac{\partial}{\partial x}\left[\sigma_{xx}\left(1 + \frac{\partial u_x}{\partial x} \right) + \sigma_{xy}\frac{\partial u_x}{\partial y} + \sigma_{xz}\frac{\partial u_x}{\partial z} \right]
$$

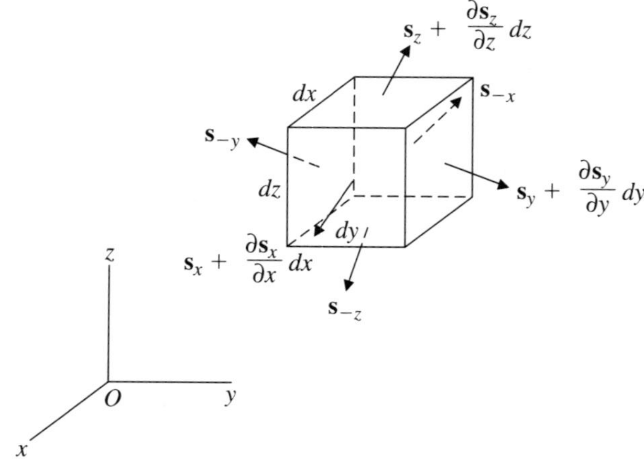

Fig. 1.2.1. Stress vectors on a volume element.

$$
+\frac{\partial}{\partial y}\left[\sigma_{yx}\left(1+\frac{\partial u_x}{\partial x}\right)+\sigma_{yy}\frac{\partial u_x}{\partial y}+\sigma_{yz}\frac{\partial u_x}{\partial z}\right]
$$

$$
+\frac{\partial}{\partial z}\left[\sigma_{zx}\left(1+\frac{\partial u_x}{\partial x}\right)+\sigma_{zy}\frac{\partial u_x}{\partial y}+\sigma_{zz}\frac{\partial u_x}{\partial z}\right]+f_x=0
$$

$$
\frac{\partial}{\partial x}\left[\sigma_{xx}\frac{\partial u_y}{\partial x}+\sigma_{xy}\left(1+\frac{\partial u_y}{\partial y}\right)+\sigma_{xz}\frac{\partial u_y}{\partial z}\right]
$$

$$
+\frac{\partial}{\partial y}\left[\sigma_{yx}\frac{\partial u_y}{\partial x}+\sigma_{yy}\left(1+\frac{\partial u_y}{\partial y}\right)+\sigma_{yz}\frac{\partial u_y}{\partial z}\right] \qquad (1.2.3)
$$

$$
+\frac{\partial}{\partial z}\left[\sigma_{zx}\frac{\partial u_y}{\partial x}+\sigma_{zy}\left(1+\frac{\partial u_y}{\partial y}\right)+\sigma_{zz}\frac{\partial u_y}{\partial z}\right]+f_y=0
$$

$$
\frac{\partial}{\partial x}\left[\sigma_{xx}\frac{\partial u_z}{\partial x}+\sigma_{xy}\frac{\partial u_z}{\partial y}+\sigma_{xz}\left(1+\frac{\partial u_z}{\partial z}\right)\right]
$$

$$
+\frac{\partial}{\partial y}\left[\sigma_{yx}\frac{\partial u_z}{\partial x}+\sigma_{yy}\frac{\partial u_z}{\partial y}+\sigma_{yz}\left(1+\frac{\partial u_z}{\partial z}\right)\right]
$$

$$
+\frac{\partial}{\partial z}\left[\sigma_{zx}\frac{\partial u_z}{\partial x}+\sigma_{zy}\frac{\partial u_z}{\partial y}+\sigma_{zz}\left(1+\frac{\partial u_z}{\partial z}\right)\right]+f_z=0.
$$

In terms of linear strains and rotations, these can be rewritten as

$$
\frac{\partial}{\partial x}\left[\sigma_{xx}(1+e_{xx})+\sigma_{xy}(e_{xy}-\omega_{xy})+\sigma_{xz}(e_{zx}+\omega_{zx})\right]
$$

$$
\frac{\partial}{\partial y}\left[\sigma_{yx}(1+e_{xx})+\sigma_{yy}(e_{xy}-\omega_{xy})+\sigma_{yz}(e_{zx}+\omega_{zx})\right]
$$

$$
\frac{\partial}{\partial z}\left[\sigma_{zx}(1+e_{xx})+\sigma_{zy}(e_{xy}-\omega_{xy})+\sigma_{zz}(e_{zx}+\omega_{zx})\right]+f_x=0
$$

$$
\frac{\partial}{\partial x}\left[\sigma_{xx}(e_{xy}+\omega_{xy})+\sigma_{xy}(1+e_{yy})+\sigma_{xz}(e_{yz}-\omega_{yz})\right]/\partial x
$$

$$+\frac{\partial}{\partial y}[\sigma_{yx}(e_{xy}+\omega_{xy})+\sigma_{yy}(1+e_{yy})+\sigma_{yz}(e_{yz}-\omega_{yz})] \qquad (1.2.4)$$

$$+\frac{\partial}{\partial z}[\sigma_{zx}(e_{xy}+\omega_{xy})+\sigma_{zy}(1+e_{yy})+\sigma_{zz}(e_{yz}-\omega_{yz})]+f_y=0$$

$$\frac{\partial}{\partial x}[\sigma_{xx}(e_{zx}-\omega_{zx})+\sigma_{xy}(e_{yz}+\omega_{yz})+\sigma_{xz}(1+e_{zz})]$$

$$+\frac{\partial}{\partial y}[\sigma_{yx}(e_{zx}-\omega_{zx})+\sigma_{yy}(e_{yz}+\omega_{yz})+\sigma_{yz}(1+e_{zz})]$$

$$+\frac{\partial}{\partial z}[\sigma_{zx}(e_{zx}-\omega_{zx})+\sigma_{zy}(e_{yz}+\omega_{yz})+\sigma_{zz}(1+e_{zz})]+f_z=0.$$

Equations (1.2.3) and (1.2.4) are the equilibrium equations for the classical nonlinear case of large finite deformations.

At a point on the exterior boundary surface of a body, either the displacement or traction may be prescribed. When the displacement is prescribed, its value in the final solution simply must be equal to the prescribed value. When the traction is prescribed, a surface element at the boundary is considered, as shown in Figure 1.2.2. The element has an outward normal $\mathbf{n}$ and is subjected to a traction

$$\mathbf{p}_n = \mathbf{i}_x p_x + \mathbf{i}_y p_y + \mathbf{i}_z p_z. \qquad (1.2.5)$$

Equilibrium of forces acting on the tetrahedron at the boundary requires that

$$\mathbf{s}_{-x}n_x + \mathbf{s}_{-y}n_y + \mathbf{s}_{-z}n_z + \mathbf{p}_n = 0, \qquad (1.2.6)$$

where $n_x = \cos(x,\mathbf{n})$, $n_y = \cos(y,\mathbf{n})$, and $n_z = \cos(z,\mathbf{n})$ are the direction cosines of $\mathbf{n}$. Since $\mathbf{s}_{-x} = -\mathbf{s}_x$, $\mathbf{s}_{-y} = -\mathbf{s}_y$, and $\mathbf{s}_{-z} = -\mathbf{s}_z$, all these stress vectors may be written in terms of the stress components according to Eqs. (1.2.1).

The traction boundary condition is thus derived by substituting Eqs. (1.2.1) and (1.2.5) in (1.2.6). The results are, in component form,

$$p_x = \left[\sigma_{xx}\left(1+\frac{\partial u_x}{\partial x}\right)+\sigma_{xy}\frac{\partial u_x}{\partial y}+\sigma_{xz}\frac{\partial u_x}{\partial z}\right]n_x$$

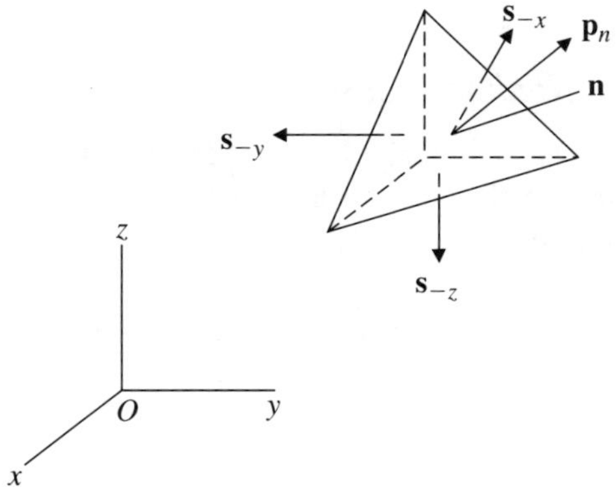

Fig. 1.2.2. Stress vectors on a surface element.

$$+ \left[\sigma_{yx} \left(1 + \frac{\partial u_x}{\partial x} \right) + \sigma_{yy} \frac{\partial u_x}{\partial y} + \sigma_{yz} \frac{\partial u_x}{\partial z} \right] n_y$$

$$+ \left[\sigma_{zx} \left(1 + \frac{\partial u_x}{\partial x} \right) + \sigma_{zy} \frac{\partial u_x}{\partial y} + \sigma_{zz} \frac{\partial u_x}{\partial z} \right] n_z$$

$$p_y = \left[\sigma_{xx} \frac{\partial u_y}{\partial x} + \sigma_{xy} \left(1 + \frac{\partial u_y}{\partial y} \right) + \sigma_{xz} \frac{\partial u_y}{\partial z} \right] n_x$$

$$+ \left[\sigma_{yx} \frac{\partial u_y}{\partial x} + \sigma_{yy} \left(1 + \frac{\partial u_y}{\partial y} \right) + \sigma_{yz} \frac{\partial u_y}{\partial z} \right] n_y$$

$$+ \left[\sigma_{zx} \frac{\partial u_y}{\partial x} + \sigma_{zy} \left(1 + \frac{\partial u_y}{\partial y} \right) + \sigma_{zz} \frac{\partial u_y}{\partial z} \right] n_z$$

$$p_z = \left[\sigma_{xx} \frac{\partial u_z}{\partial x} + \sigma_{xy} \frac{\partial u_z}{\partial y} + \sigma_{xz} \left(1 + \frac{\partial u_z}{\partial z} \right) \right] n_x$$

$$+ \left[\sigma_{yx} \frac{\partial u_z}{\partial x} + \sigma_{yy} \frac{\partial u_z}{\partial y} + \sigma_{yz} \left(1 + \frac{\partial u_z}{\partial z} \right) \right] n_y$$

$$+ \left[\sigma_{zx} \frac{\partial u_z}{\partial x} + \sigma_{zy} \frac{\partial u_z}{\partial y} + \sigma_{zz} \left(1 + \frac{\partial u_z}{\partial z} \right) \right] n_z, \tag{1.2.7}$$

which are for the classical nonlinear case. In terms of linear strains and rotations, these are rewritten as

$$p_x = [\sigma_{xx}(1 + e_{xx}) + \sigma_{xy}(e_{xy} - \omega_{xy}) + \sigma_{xz}(e_{zx} + \omega_{zx})]n_x$$
$$+ [\sigma_{yx}(1 + e_{xx}) + \sigma_{yy}(e_{xy} - \omega_{xy}) + \sigma_{yz}(e_{zx} + \omega_{zx})]n_y$$
$$+ [\sigma_{zx}(1 + e_{xx}) + \sigma_{zy}(e_{xy} - \omega_{xy}) + \sigma_{zz}(e_{zx} + \omega_{zx})]n_z$$
$$p_y = [\sigma_{xx}(e_{xy} + \omega_{xy}) + \sigma_{xy}(1 + e_{yy}) + \sigma_{xz}(e_{yz} - \omega_{yz})]n_x$$
$$+ [\sigma_{yx}(e_{xy} + \omega_{xy}) + \sigma_{yy}(1 + e_{yy}) + \sigma_{yz}(e_{yz} - \omega_{yz})]n_y \tag{1.2.8}$$
$$+ [\sigma_{zx}(e_{xy} + \omega_{xy}) + \sigma_{zy}(1 + e_{yy}) + \sigma_{zz}(e_{yz} - \omega_{yz})]n_z$$
$$p_z = [\sigma_{xx}(e_{zx} - \omega_{zx}) + \sigma_{xy}(e_{yz} + \omega_{yz}) + \sigma_{xz}(1 + e_{zz})]n_x$$
$$+ [\sigma_{yx}(e_{zx} - \omega_{zx}) + \sigma_{yy}(e_{yz} + \omega_{yz}) + \sigma_{yz}(1 + e_{zz})]n_y$$
$$+ [\sigma_{zx}(e_{zx} - \omega_{zx}) + \sigma_{zy}(e_{yz} + \omega_{yz}) + \sigma_{zz}(1 + e_{zz})]n_z.$$

1.2.3 Simplified Nonlinear Cases

If the linear strains are not only negligibly small compared with unity but also much smaller than the rotations, they may be dropped from Eqs. (1.2.4). The equilibrium equations then reduce to

$$\frac{\partial}{\partial x}(\sigma_{xx} - \sigma_{xy}\omega_{xy} + \sigma_{xz}\omega_{zx}) + \frac{\partial}{\partial y}(\sigma_{yx} - \sigma_{yy}\omega_{xy} + \sigma_{yz}\omega_{zx})$$

$$+ \frac{\partial}{\partial z}(\sigma_{zx} - \sigma_{zy}\omega_{xy} + \sigma_{zz}\omega_{zx}) + f_x = 0$$

$$\frac{\partial}{\partial x}(\sigma_{xx}\omega_{xy} + \sigma_{xy} - \sigma_{xz}\omega_{yz}) + \frac{\partial}{\partial y}(\sigma_{yx}\omega_{xy} + \sigma_{yy} - \sigma_{yz}\omega_{yz})$$

$$+\frac{\partial}{\partial z}(\sigma_{zx}\omega_{xy} + \sigma_{zy} - \sigma_{zz}\omega_{yz}) + f_y = 0 \qquad (1.2.9)$$

$$\frac{\partial}{\partial x}(-\sigma_{xx}\omega_{zx} + \sigma_{xy}\omega_{yz} + \sigma_{xz}) + \frac{\partial}{\partial y}(-\sigma_{yx}\omega_{zx} + \sigma_{yy}\omega_{yz} + \sigma_{yz})$$

$$+\frac{\partial}{\partial z}(-\sigma_{zx}\omega_{zx} + \sigma_{zy}\omega_{yz} + \sigma_{zz}) + f_z = 0.$$

These are for the simplified nonlinear case of small strains and large rotations (Novozhilov 1948).

For application to thin structures such as beams, plates, and shells, a new simplification has been proposed recently (Yu 1991, 1995a,b). Thus, in the case of a plate, with the middle plane chosen in the xy-directions and thickness in the z-direction, the nonlinear effects are neglected in the first two of Eqs. (1.2.9) and retained only in the third equation. Equations (1.2.9) then further reduce to

$$\frac{\partial\sigma_{xx}}{\partial x} + \frac{\partial\sigma_{yx}}{\partial y} + \frac{\partial\sigma_{zx}}{\partial z} + f_x = 0$$

$$\frac{\partial\sigma_{xy}}{\partial x} + \frac{\partial\sigma_{yy}}{\partial y} + \frac{\partial\sigma_{zy}}{\partial z} + f_y = 0$$

$$\frac{\partial}{\partial x}(-\sigma_{xx}\omega_{zx} + \sigma_{xy}\omega_{yz} + \sigma_{xz}) + \frac{\partial}{\partial y}(-\sigma_{yx}\omega_{zx} + \sigma_{yy}\omega_{yz} + \sigma_{yz}) \qquad (1.2.10)$$

$$+\frac{\partial}{\partial z}(-\sigma_{zx}\omega_{zx} + \sigma_{zy}\omega_{yz} + \sigma_{zz}) + f_z = 0.$$

These are the equilibrium equations for the new simplified nonlinear case. As indicated in the preceding section, both of the simplified nonlinear cases are characterized by small finite deformations.

For the simplified nonlinear case, the traction boundary conditions in Eqs. (1.2.8) reduce to

$$p_x = (\sigma_{xx} - \sigma_{xy}\omega_{xy} + \sigma_{xz}\omega_{zx})n_x$$
$$+(\sigma_{yx} - \sigma_{yy}\omega_{xy} + \sigma_{yz}\omega_{zx})n_y$$
$$+(\sigma_{zx} - \sigma_{zy}\omega_{xy} + \sigma_{zz}\omega_{zx})n_z$$
$$p_y = (\sigma_{xx}\omega_{xy} + \sigma_{xy} - \sigma_{xz}\omega_{yz})n_x$$
$$+(\sigma_{yx}\omega_{xy} + \sigma_{yy} - \sigma_{yz}\omega_{yz})n_y \qquad (1.2.11)$$
$$+(\sigma_{zx}\omega_{xy} + \sigma_{zy} - \sigma_{zz}\omega_{yz})n_z$$
$$p_z = (-\sigma_{xx}\omega_{zx} + \sigma_{xy}\omega_{yz} + \sigma_{xz})n_x$$
$$+(-\sigma_{yx}\omega_{zx} + \sigma_{yy}\omega_{yz} + \sigma_{yz})n_y$$
$$+(-\sigma_{zx}\omega_{zx} + \sigma_{zy}\omega_{yz} + \sigma_{zz})n_z.$$

For the new simplified nonlinear case, Eqs. (1.2.11) further reduce to

$$p_x = \sigma_{xx}n_x + \sigma_{yx}n_y + \sigma_{xz}n_z$$
$$p_y = \sigma_{xy}n_x + \sigma_{yy}n_y + \sigma_{zy}n_z$$

$$p_z = (-\sigma_{xx}\omega_{zx} + \sigma_{xy}\omega_{yz} + \sigma_{xz})n_x \qquad (1.2.12)$$
$$+(-\sigma_{yx}\omega_{zx} + \sigma_{yy}\omega_{yz} + \sigma_{yz})n_y$$
$$+(-\sigma_{zx}\omega_{zx} + \sigma_{zy}\omega_{yz} + \sigma_{zz})n_z.$$

1.2.4 Linear Case

When all nonlinear terms are neglected from the above results, the familiar results in linear elasticity are obtained. Thus, the equilibrium equations become

$$\frac{\partial \sigma_{xx}}{\partial x} + \frac{\partial \sigma_{yx}}{\partial y} + \frac{\partial \sigma_{zx}}{\partial z} + f_x = 0$$
$$\frac{\partial \sigma_{xy}}{\partial x} + \frac{\partial \sigma_{yy}}{\partial y} + \frac{\partial \sigma_{zy}}{\partial z} + f_y = 0 \qquad (1.2.13)$$
$$\frac{\partial \sigma_{xz}}{\partial x} + \frac{\partial \sigma_{yz}}{\partial y} + \frac{\partial \sigma_{zz}}{\partial z} + f_z = 0$$

and the traction boundary conditions become

$$p_x = \sigma_{xx}n_x + \sigma_{yx}n_y + \sigma_{zx}n_z$$
$$p_y = \sigma_{xy}n_x + \sigma_{yy}n_y + \sigma_{zy}n_z \qquad (1.2.14)$$
$$p_z = \sigma_{xz}n_x + \sigma_{yz}n_y + \sigma_{zz}n_z.$$

As was mentioned in the preceding section, these are characterized by infinitesimal deformations.

1.3 Strain Energy Function and Principle of Virtual Work

The strain energy function is an important concept that has been much discussed in the literature. Early discussion on the subject, for small deformations in the dynamic case, may be found in the classical treatise on elasticity by Love (1927). Love started with the laws of thermodynamics and showed the existence of the strain energy function for the adiabatic and isothermal cases. He then associated the strain energy function with Hooke's law. In his mathematical treatise, he even included a section on the indirectness of experimental results, as he realized that stress and strain components inside a solid could not be measured directly. A clear discussion of the strain energy function has also been given in the standard text by Fung (1965).

In this section, we start with an evaluation of the virtual work, from which the principle of virtual work is then derived for the classical nonlinear case of large finite deformations. In this discussion, the intimate relations between the components of Kirchhoff's stress tensor and those of Green's strain tensor, together with the concept of the strain energy function, will emerge as natural consequences.

Consider an elastic body in equilibrium under the infuence of the body force

$$\mathbf{f} = \mathbf{i}_x f_x + \mathbf{i}_y f_y + \mathbf{i}_z f_z$$

over the volume V of the body, and the surface traction

$$\mathbf{p} = \mathbf{i}_x p_x + \mathbf{i}_y p_y + \mathbf{i}_z p_z$$

over the part of the surface of the body designated by S_p. We assign to the body a virtual displacement

$$\delta \mathbf{u} = \mathbf{i}_x \delta u_x + \mathbf{i}_y \delta u_y + \mathbf{i}_z \delta u_z$$

Since δu must vanish over that part of the surface of the body, designated by S_u, on which the displacement is prescribed, S_u need not be considered in the evaluation of the virtual work. The virtual work is then, by definition,

$$\delta W = \int_V \mathbf{f} \cdot \delta \mathbf{u} \, dV + \int_{S_p} \mathbf{p} \cdot \delta \mathbf{u} \, dS. \tag{1.3.1}$$

By virtue of the equilibrium equation (1.2.2), the volume integral in Eq. (1.3.1) may be rewritten as

$$\int_V \mathbf{f} \cdot \delta \mathbf{u} \, dV = - \int_V \left(\frac{\partial \mathbf{s}_x}{\partial x} + \frac{\partial \mathbf{s}_y}{\partial y} + \frac{\partial \mathbf{s}_z}{\partial z} \right) \cdot \delta \mathbf{u} \, dV. \tag{1.3.2}$$

Similarly, by virtue of the tration boundary conditions (1.2.7), the surface integral in Eq. (1.3.1) becomes

$$\int_{S_p} \mathbf{p} \cdot \delta \mathbf{u} \, dS = \int_{S_p} (p_x \delta u_x + p_y \delta u_y + p_z \delta u_z) \, dS$$

$$= \int_{S_p} \left[\left\{ \left[\sigma_{xx} \left(1 + \frac{\partial u_x}{\partial x} \right) + \sigma_{xy} \frac{\partial u_x}{\partial y} + \sigma_{xz} \frac{\partial u_x}{\partial z} \right] \delta u_x \right. \right.$$

$$+ \left[\sigma_{xx} \frac{\partial u_y}{\partial x} + \sigma_{xy} \left(1 + \frac{\partial u_y}{\partial y} \right) + \sigma_{xz} \frac{\partial u_y}{\partial z} \right] \delta u_y$$

$$+ \left[\sigma_{xx} \frac{\partial u_z}{\partial x} + \sigma_{xy} \frac{\partial u_z}{\partial y} + \sigma_{xz} \left(1 + \frac{\partial u_z}{\partial z} \right) \right] \delta u_z \right\} n_x$$

$$+ \left\{ \ldots \right\} n_y + \left\{ \ldots \right\} n_z \right] dS,$$

which is then transformed into a volume integral by means of Gauss' theorem. By further introducing $\mathbf{s}_x$, $\mathbf{s}_y$, and $\mathbf{s}_z$ from Eqs. (1.2.1), the result takes the form

$$\int_{S_p} \mathbf{p} \cdot \delta \mathbf{u} \, dS = \int_V \left[\frac{\partial (\mathbf{s}_x \cdot \delta \mathbf{u})}{\partial x} + \frac{\partial (\mathbf{s}_y \cdot \delta \mathbf{u})}{\partial y} + \frac{\partial (\mathbf{s}_z \cdot \delta \mathbf{u})}{\partial z} \right] dV. \tag{1.3.3}$$

Substitution of Eqs. (1.3.2) and (1.3.3) in (1.3.1) now yields the virtual work

$$\delta W = \int_V \left[\mathbf{s}_x \cdot \frac{\partial (\delta \mathbf{u})}{\partial x} + \mathbf{s}_y \cdot \frac{\partial (\delta \mathbf{u})}{\partial y} + \mathbf{s}_z \cdot \frac{\partial (\delta \mathbf{u})}{\partial z} \right] dV. \tag{1.3.4}$$

Next, according to Eqs. (1.1.3) and (1.1.5), we have

$$\frac{\partial(\delta\mathbf{u})}{\partial x} = \delta\mathbf{g}_x, \ldots$$

and

$$\delta\epsilon_{xx} = \mathbf{g}_x \cdot \delta\mathbf{g}_x, \ldots, \quad \delta\epsilon_{xy} = \tfrac{1}{2}(\mathbf{g}_x \cdot \delta\mathbf{g}_y + \mathbf{g}_y \cdot \delta\mathbf{g}_x), \ldots.$$

By making use of these results together with Eqs. (1.2.1), the virtual work in Eq. (1.3.4) becomes

$$\begin{aligned}
\delta W = \int_V (&\sigma_{xx}\delta\epsilon_{xx} + \sigma_{xy}\delta\epsilon_{xy} + \sigma_{xz}\delta\epsilon_{xz} \\
&+ \sigma_{yx}\delta\epsilon_{yx} + \sigma_{yy}\delta\epsilon_{yy} + \sigma_{yz}\delta\epsilon_{yz} \\
&+ \sigma_{zx}\delta\epsilon_{zx} + \sigma_{zy}\delta\epsilon_{zy} + \sigma_{zz}\delta\epsilon_{zz})\, dV.
\end{aligned} \tag{1.3.5}$$

Equation (1.3.5) is an important result in that each of the components of Kirchhoff's stress tensor is associated with the corresponding component of Green's strain tensor. In fact, as a virtual displacement is assigned, each of the products in the integral represents the increment of strain energy per unit volume generated by a stress component with the increment of the corresponding strain component. A strain energy function U_0 can thus be introduced in such a way that

$$\sigma_{xx} = \frac{\partial U_0}{\partial \epsilon_{xx}}, \ldots, \quad \sigma_{xy} = \frac{\partial U_0}{\partial \epsilon_{xy}}, \ldots. \tag{1.3.6}$$

If the strain energy function exists, the integrand in Eq. (1.3.5) becomes an exact differential since

$$\begin{aligned}
\sigma_{xx}&\delta\epsilon_{xx} + \sigma_{xy}\delta\epsilon_{xy} + \sigma_{xz}\delta\epsilon_{xz} \\
&+ \sigma_{yx}\delta\epsilon_{yx} + \sigma_{yy}\delta\epsilon_{yy} + \sigma_{yz}\delta\epsilon_{yz} \\
&+ \sigma_{zx}\delta\epsilon_{zx} + \sigma_{zy}\delta\epsilon_{zyx} + \sigma_{zz}\delta\epsilon_{zz} \\
= \frac{\partial U_0}{\partial \epsilon_{xx}}&\delta\epsilon_{xx} + \frac{\partial U_0}{\partial \epsilon_{xy}}\delta\epsilon_{xy} + \frac{\partial U_0}{\partial \epsilon_{xz}}\delta\epsilon_{xz} \\
&+ \frac{\partial U_0}{\partial \epsilon_{yx}}\delta\epsilon_{yx} + \frac{\partial U_0}{\partial \epsilon_{yy}}\delta\epsilon_{yy} + \frac{\partial U_0}{\partial \epsilon_{yz}}\delta\epsilon_{yz} \\
&+ \frac{\partial U_0}{\partial \epsilon_{zx}}\delta\epsilon_{zx} + \frac{\partial U_0}{\partial \epsilon_{zy}}\delta\epsilon_{zy} + \frac{\partial U_0}{\partial \epsilon_{zz}}\delta\epsilon_{zz} \\
= \delta U_0&.
\end{aligned} \tag{1.3.7}$$

Through integration we introduce

$$\delta U = \int_V \delta U_0\, dV. \tag{1.3.8}$$

Equation (1.3.5) then may be written in the final simple form

$$\delta W = \delta U, \tag{1.3.9}$$

which is the principle of virtual work for an elastic body undergoing large finite deformations. It states that, as a virtual displacement is assigned, the increment of strain energy in the body is equal to the virtual work done by all external forces acting on the body. For a rigid body, the principle reduces to the usual simple form of

$$\delta W = 0$$

We note that the derivation of Eq. (1.3.9) depends on not only the equilibrium equations and traction boundary conditions, but also on the existence of a strain energy function. The latter provides an important link with the elastic stress–strain relations. When a solid is strained within the elastic limit and obeys the generalized Hook's law, each of the stress components at any point is a linear function of the strain components. Just as in linear elasticity, the strain energy function is then a homogeneous quadratic function of the strain components, although the strains are now nonlinear functions of displacement gradients. Elastic stress–strain relations will be discussed in a later section.

1.4   Hamilton's Principle and Variational Equations of Motion

1.4.1   Hamilton's Principle

Hamilton's principle is an extension of the principle of virtual work from statics to dynamics. It can be formulated for either a rigid or deformable body by invoking D'Alembert's principle to accommodate inertia forces. Integration with respect to time is carried out between fixed initial and final instants of time t_0 and t_1, under the constraint that the virtual displacement is required to vanish at t_0 and t_1. In the case of a deformable body, the virtual displacement must also vanish at those parts of the body on which displacement is prescribed.

Let δW_u denote the virtual work done by the inertia forces. Then, upon integration with respect to t,

$$\int_{t_0}^{t_1} \delta W_u \, dt = -\int_V \int \rho \frac{\partial^2 \mathbf{u}}{\partial t^2} \cdot \delta \mathbf{u} \, dt \, dV$$

$$= -\int_V \rho \frac{\partial \mathbf{u}}{\partial t} \cdot \delta \mathbf{u} \, dV \Big|_{t_0}^{t_1} + \int_{t_0}^{t_1} dt \int_V \rho \frac{\partial \mathbf{u}}{\partial t} \cdot \frac{\partial (\delta \mathbf{u})}{\partial t} \, dV \quad (1.4.1)$$

$$= 0 + \int_{t_0}^{t_1} \delta T \, dt,$$

where ρ is the density of the solid in the undeformed state. In the last line of this result, the first term becomes zero since $\delta \mathbf{u}$ vanishes at t_0 and t_1, and the second term becomes associated with the variation of the total kinetic energy T:

$$\delta T = \delta \int_V \frac{1}{2} \rho \frac{\partial \mathbf{u}}{\partial t} \cdot \frac{\partial \mathbf{u}}{\partial t} \, dV. \qquad (1.4.2)$$

The principle of virtual work in Eq. (1.3.9) is now integrated with respect to time similarly and written in the form

$$\int_{t_0}^{t_1} (\delta W - \delta U)\, dt = 0.$$

By introducing δW_u as an additional part to account for the inertia forces, this is next extended to the dynamic case:

$$\int_{t_0}^{t_1} (\delta W_u + \delta W - \delta U)\, dt = 0.$$

Replacing δW_u by δT according to Eq. (1.4.1), this then becomes, finally,

$$\delta \int_{t_0}^{t_1} L\, dt = \delta \int_{t_0}^{t_1} (T - U + W)\, dt = 0, \tag{1.4.3}$$

where $L = T - U + W$ is the Lagrangian function.

Equation (1.4.3) is Hamilton's principle for a solid undergoing large finite deformations. As a variational principle, it states that the variation of the integral of the Lagrangian function over the time interval between t_0 and t_1 vanishes or, equivalently, the value of the integral is stationary, provided that the variations of the displacements vanish not only at t_0 and t_1, but also at those parts of the boundary where displacements are prescribed.

1.4.2 Variational Equation of Motion in Classical Nonlinear Case

We recall that, in Section 1.3, the principle of virtual work was derived from the equilibrium equation and traction boundary condition. In the dynamic case, Hamilton's principle may similarly be derived, since it is only necessary to replace the equilibrium equation (1.2.2) by the equation of motion,

$$\frac{\partial s_x}{\partial x} + \frac{\partial s_y}{\partial y} + \frac{\partial s_z}{\partial z} + \mathbf{f} = \rho \frac{\partial^2 \mathbf{u}}{\partial t^2}. \tag{1.4.4}$$

However, we shall not pursue this parallel case in dynamics. Instead, the converse situation will be considered here. This is to accept Hamilton's principle in Eq. (1.4.3) as a starting point. By carrying out variations according to the variational principle, a variational equation of motion will be derived as the end result. According to Love (1927), the variational equation of motion was first derived from Hamilton's principle in linear elasticity by Kirchhoff (1883).

In the classical nonlinear case of large finite deformations, δT is given by Eq. (1.4.2), δU is given by Eqs. (1.3.7) and (1.3.8), and δW is given by Eq. (1.3.1). By substituting these into Eq. (1.4.3) and carrying out the variations, the end result is readily shown to be

$$\int_{t_0}^{t_1} dt \int_V \left(\left\{ \frac{\partial}{\partial x} \left[\sigma_{xx} \left(1 + \frac{\partial u_x}{\partial x} \right) + \sigma_{xy} \frac{\partial u_x}{\partial y} + \sigma_{xz} \frac{\partial u_x}{\partial z} \right] \right. \right.$$

$$+ \frac{\partial}{\partial y}\left[\sigma_{yx}\left(1 + \frac{\partial u_x}{\partial x}\right) + \sigma_{yy}\frac{\partial u_x}{\partial y} + \sigma_{yz}\frac{\partial u_x}{\partial z}\right]$$

$$+ \frac{\partial}{\partial z}\left[\sigma_{zx}\left(1 + \frac{\partial u_x}{\partial x}\right) + \sigma_{zy}\frac{\partial u_x}{\partial y} + \sigma_{zz}\frac{\partial u_x}{\partial z}\right]$$

$$+ f_x - \rho\frac{\partial^2 u_x}{\partial t^2}\Big\}\delta u_x + \{\ldots\}\delta u_y + \{\ldots\}\delta u_z\Big)\,dV$$

$$- \int_{t_0}^{t_1} dt \int_{S_p}\left(\Big\{\left[\sigma_{xx}\left(1 + \frac{\partial u_x}{\partial x}\right) + \sigma_{xy}\frac{\partial u_x}{\partial y} + \sigma_{xz}\frac{\partial u_x}{\partial z}\right]n_x\right.$$

$$+ \left[\sigma_{yx}\left(1 + \frac{\partial u_x}{\partial x}\right) + \sigma_{yy}\frac{\partial u_x}{\partial y} + \sigma_{yz}\frac{\partial u_x}{\partial z}\right]n_y$$

$$+ \left[\sigma_{zx}\left(1 + \frac{\partial u_x}{\partial x}\right) + \sigma_{zy}\frac{\partial u_x}{\partial y} + \sigma_{zz}\frac{\partial u_x}{\partial z}\right]n_z$$

$$\left. - p_x\Big\}\delta u_x + \{\ldots\}\delta u_y + \{\ldots\}\delta u_z\right)\,dS = 0. \qquad (1.4.5)$$

In terms of linear strains and rotations, this appears in the form

$$\int_{t_0}^{t_1} dt \int_V \Big(\Big\{\frac{\partial}{\partial x}\left[\sigma_{xx}(1 + e_{xx}) + \sigma_{xy}(e_{xy} - \omega_{xy}) + \sigma_{xz}(e_{xz} + \omega_{zx})\right]$$

$$+ \frac{\partial}{\partial z}\left[\sigma_{yx}(1 + e_{xx}) + \sigma_{yy}(e_{xy} - \omega_{xy}) + \sigma_{zx}(e_{zx} + \omega_{zx})\right]$$

$$+ \frac{\partial}{\partial z}\left[\sigma_{zx}(1 + e_{xx}) + \sigma_{zy})e_{xy} - \omega_{xy} - \omega_{xy}) + \sigma_{zx} + \omega_{zx})\right]$$

$$+ f_x - \rho\frac{\partial^2 u_x}{\partial t^2}\Big\}\delta u_x + \{\ldots\}\delta u_y + \{\ldots\}\delta u_z\Big)\,dV$$

$$- \int_{t_0}^{t_1} dt \int_{S_p}\Big(\{[\sigma_{xx}(1 + e_{xx}) + \sigma_{xy}(e_{xy} - \omega_{xy}) + \sigma_{xz})(e_{zx} + \omega_{zx})]n_x$$

$$+ [\sigma_{yx}(1 + e_{xx}) + \sigma_{yy}(e_{xy} - \omega_{xy}) + \sigma_{yz}(e_{zx} + \omega_{zx})]n_y$$

$$+ [\sigma_{zx}(1 + e_{xx}) + \sigma_{zy}(e_{xy} - \omega_{xy}) + \sigma_{zz}(e_{zx} + \omega_{zx})]n_z$$

$$- p_x\}\delta u_x + \{\ldots\}\delta u_y + \{\ldots\}\delta u_z\Big)\,dS = 0 \qquad (1.4.6)$$

Equation (1.4.5) or (1.4.6) is the variational equation of motion from which the stress equations of motion and traction boundary conditions can be written as the Euler equations according to the rules of calculus of variations. The static counterpart of these results was discussed in Section 1.2.

As already pointed out, the starting point here is Hamilton's principle, which is a variational principle, and the end result is the variational equation of motion. For the classical nonlinear case of large finite deformations under consideration, the variational principle and variational equation of motion correspond to each other exactly, and we may start from either of the two and deduce the other.

1.4.3 Variational Equations of Motion in Simplified Nonlinear Cases

In an attempt to formulate similar variational equations of motion for the two simplified nonlinear cases, we start with the variational equation of motion (1.4.5) for the general classical nonlinear case. The integrands in the equation are simplified in the same manner as the stress equations and traction boundary conditions were treated in Section 1.2, by first writing these in terms of the linear strains and rotations and then neglecting some of the terms. Thus, we obtain for the first simplified nonlinear case

$$
\int_{t_0}^{t_1} dt \int_V \left\{ \left[\frac{\partial}{\partial x}(\sigma_{xx} - \sigma_{xy}\omega_{xy} + \sigma_{xz}\omega_{zx}) \right. \right.
$$

$$
+ \frac{\partial}{\partial y}(\sigma_{yx} - \sigma_{yy}\omega_{xy} + \sigma_{yz}\omega_{zx})
$$

$$
+ \frac{\partial}{\partial z}(\sigma_{zx} - \sigma_{zy}\omega_{xy} + \sigma_{zz}\omega_{zx})
$$

$$
\left. + f_x - \rho \frac{\partial^2 u_x}{\partial t^2} \right] \delta u_x + [\ldots]\delta u_y + [\ldots]\delta u_z \right\} dV
$$

$$
- \int_{t_0}^{t_1} dt \int_{S_p} \left\{ [(\sigma_{xx} - \sigma_{xy}\omega_{xy} + \sigma_{xz}\omega_{zx})n_x \right.
$$

$$
+ (\sigma_{yx} - \sigma_{yy}\omega_{xy} + \sigma_{yz}\omega_{zx})n_y
$$

$$
+ (\sigma_{zx} - \sigma_{zy}\omega_{xy} + \sigma_{zz}\omega_{zx})n_z
$$

$$
\left. - p_x]\delta u_x + [\ldots]\delta u_y + [\ldots]\delta u_z \right\} dS = 0. \tag{1.4.7}
$$

Similarly, for the second and new simplified nonlinear case,

$$
\int_{t_0}^{t_1} dt \int_V \left\{ \left[\frac{\partial \sigma_{xx}}{\partial x} + \frac{\partial \sigma_{yx}}{\partial y} + \frac{\partial \sigma_{zx}}{\partial z} + f_x - \rho \ddot{u}_x \right] \delta u_x \right.
$$

$$
+ \left[\frac{\partial \sigma_{xy}}{\partial x} + \frac{\partial \sigma_{yy}}{\partial y} + \frac{\partial \sigma_{zy}}{\partial z} + f_y - \rho \ddot{u}_y \right] \delta u_y
$$

$$
+ \left[\frac{\partial}{\partial x}(\sigma_{xz} - \sigma_{xx}\omega_{zx} + \sigma_{xy}\omega_{yz}) \right.
$$

$$
+ \frac{\partial}{\partial y}(\sigma_{yz} - \sigma_{yx}\omega_{zx} + \sigma_{yy}\omega_{yz})
$$

$$
\left. \left. + \frac{\partial}{\partial z}(\sigma_{zz} - \sigma_{zx}\omega_{zx} + \sigma_{zy}\omega_{yz}) + f_z - \rho \ddot{u}_z \right] \delta u_z \right\} dV
$$

$$
- \int_{t_0}^{t_1} dt \int_{S_p} \left\{ [\sigma_{xx}n_x + \sigma_{yx}n_y + \sigma_{zx}n_z - p_x]\delta u_x \right.
$$

$$
+ [\sigma_{xy}n_x + \sigma_{yy}n_y + \sigma_{zy}n_z - p_y]\delta u_y
$$

$$
+ [(\sigma_{xz} - \sigma_{xx}\omega_{zx} + \sigma_{xy}\omega_{yz})n_x
$$

$$+(\sigma_{yz} - \sigma_{yx}\omega_{zx} + \sigma_{yy}\omega_{yz})n_y$$

$$+(\sigma_{zz} - \sigma_{zx}\omega_{zx} + \sigma_{zy}\omega_{yz})n_z - p_z]\delta u_z\} \, dS = 0. \qquad (1.4.8)$$

At this point, we note that by introducing the simplifications adopted for the two simplified nonlinear cases in Section 1.2, a strain energy function no longer exists, as will be shown in the next section. This means that Eqs. (1.4.7) and (1.4.8) cannot be derived exactly from a variational principle such as Hamilton's principle. They are therefore not true variational equations, and we propose to call them pseudo-variational equations of motion (Yu 1991, 1995a,b).

1.4.4 Variational Equation of Motion in Linear Case

For future reference, we record here the linear version of the variational equation of motion. By neglecting the nonlinear terms, Eqs. (1.4.5) and (1.4.6) reduce to

$$\int_{t_0}^{t_1} dt \int_V \left\{ \left[\frac{\partial \sigma_{xx}}{\partial x} + \frac{\partial \sigma_{yx}}{\partial y} + \frac{\partial \sigma_{zx}}{\partial z} + f_x - \rho \ddot{u}_x \right] \delta u_x \right.$$

$$+ \left[\frac{\partial \sigma_{xy}}{\partial x} + \frac{\partial \sigma_{yy}}{\partial y} + \frac{\partial \sigma_{zy}}{\partial z} + f_y - \rho \ddot{u}_y \right] \delta u_y$$

$$\left. + \left[\frac{\partial \sigma_{xz}}{\partial x} + \frac{\partial \sigma_{yz}}{\partial y} + \frac{\partial \sigma_{zz}}{\partial z} + f_z - \rho \ddot{u}_z \right] \delta u_z \right\} \, dV$$

$$- \int_{t_0}^{t_1} dt \int_{S_p} \{ [\sigma_{xx}n_x + \sigma_{yx}n_y + \sigma_{zx}n_z - p_x] \delta u_x$$

$$+ [\sigma_{xy}n_x + \sigma_{yy}n_y + \sigma_{zy}n_z - p_y] \delta u_y$$

$$+ [\sigma_{xz}n_x + \sigma_{yz}n_y + \sigma_{zz}n_z - p_z] \delta u_z\} \, dS = 0, \qquad (1.4.9)$$

which also can be deduced from Eqs. (1.4.7) and (1.4.8) but is now, again, a true variational equation of motion.

1.5 Pseudo-Variational Equations of Motion

In this section, we derive once more a variational equation of motion from Hamilton's principle by starting with a general functional form of the nonlinear strain, expressed in terms of the linear strains and rotations, as shown by Yu (1964). The general results are then applied to the reexamination of the classical nonlinear case and the two simplified nonlinear cases. Things will work out perfectly with the classical case, but not with the simplified cases. To accommodate the latter situation, the name pseudo-variational equation of motion has been proposed.

1.5.1 A General Form of Variational Equation of Motion

Hamilton's principle has the same form as before:

$$\delta \int_{t_0}^{t_1} L\,dt = \delta \int_{t_0}^{t_1} (T - U + W)\,dt = 0 \qquad (1.5.1)$$

with

$$\delta \int_{t_0}^{t_1} T\,dt = \int_V \rho \dot{u}_i \delta u_i\,dV \Big|_{t_0}^{t_1} - \int_{t_0}^{t_1} dt \int_V \rho \ddot{u}_i \delta u_i\,dV = 0 - \cdots \qquad (1.5.2)$$

$$\delta \int_{t_0}^{t_1} U\,dt = \int_{t_0}^{t_1} dt \int_V \sigma_{ij} \delta \varepsilon_{ij}\,dV \qquad (1.5.3)$$

$$\delta \int_{t_0}^{t_1} W\,dt = \int_{t_0}^{t_1} \left[\int_V f_i \delta u_i\,dV + \int_{S_p} p_i \delta u_i\,dS \right] dt. \qquad (1.5.4)$$

Since ε_{ij} are functions of the first derivatives of displacements, they always may be expressed in terms of the linear strains and rotations; thus,

$$\varepsilon_{ij} = \varepsilon_{ij}(e_{mn}, \omega_{mn}) \quad (m, n = 1, 2, 3), \qquad (1.5.5)$$

where

$$e_{mn} = \tfrac{1}{2}(u_{m,n} + u_{n,m}), \quad \omega_{mn} = \tfrac{1}{2}(u_{m,n} - u_{n,m}). \qquad (1.5.6)$$

and a comma followed by a subscript denotes differentiation with respect to the corresponding coordinate. Introducing e_{mn} and ω_{mn} from Eqs. (1.5.6) and making use of Gauss' theorem, we find, as needed in Eq. (1.5.3),

$$
\begin{aligned}
\int_V \sigma_{ij} \delta \varepsilon_{ij}\,dV = \int_{S_p} \frac{1}{2} \Bigg[& \sigma_{ij} \left(\frac{\partial \varepsilon_{ij}}{\partial e_{mn}} + \frac{\partial \varepsilon_{ij}}{\partial \omega_{mn}} \right) \nu_n \delta_{\ell m} \\
& + \sigma_{ij} \left(\frac{\partial \varepsilon_{ij}}{\partial e_{mn}} - \frac{\partial \varepsilon_{ij}}{\partial \omega_{mn}} \right) \nu_m \delta_{\ell n} \Bigg] \delta u_\ell\,dS \\
& - \int_V \frac{1}{2} \Bigg\{ \left[\sigma_{ij} \left(\frac{\partial \varepsilon_{ij}}{\partial e_{mn}} + \frac{\partial \varepsilon_{ij}}{\partial \omega_{mn}} \right) \right]_{,n} \delta_{\ell m} \\
& + \left[\sigma_{ij} \left(\frac{\partial \varepsilon_{ij}}{\partial e_{mn}} - \frac{\partial \varepsilon_{ij}}{\partial \omega_{mn}} \right) \right]_{,m} \delta_{\ell n} \Bigg\} \delta u_\ell\,dV, \qquad (1.5.7)
\end{aligned}
$$

in which $\delta_{\ell m}$ is the Kronecker delta and $\nu_n = \cos(\nu, n)$ is the direction cosine. By virtue of Eqs. (1.5.2) through (1.5.7), Eq. (1.5.1) becomes

$$
\begin{aligned}
\int_{t_0}^{t_1} dt \int_V \Bigg\{ & \frac{1}{2} \left[\sigma_{ij} \left(\frac{\partial \varepsilon_{ij}}{\partial e_{mn}} + \frac{\partial \varepsilon_{ij}}{\partial \omega_{mn}} \right) \right]_{,n} \delta_{\ell m} \\
& + \frac{1}{2} \left[\sigma_{ij} \left(\frac{\partial \varepsilon_{ij}}{\partial e_{mn}} - \frac{\partial \varepsilon_{ij}}{\partial \omega_{mn}} \right) \right]_{,n} \delta_{\ell n} + f_\ell - \rho \ddot{u}_\ell \Bigg\} \delta u_\ell\,dV \\
- \int_{t_0}^{t_1} dt \int_{S_p} \Bigg[& \frac{1}{2} \sigma_{ij} \left(\frac{\partial \varepsilon_{ij}}{\partial e_{mn}} + \frac{\partial \varepsilon_{ij}}{\partial \omega_{mn}} \right) \nu_n \delta_{\ell m} \\
& + \frac{1}{2} \sigma_{ij} \left(\frac{\partial \varepsilon_{ij}}{\partial e_{mn}} - \frac{\partial \varepsilon_{ij}}{\partial \omega_{mn}} \right) \nu_m \delta_{\ell n} - p_\ell \Bigg] \delta u_\ell\,dS = 0, \qquad (1.5.8)
\end{aligned}
$$

which is the variational equation of motion in a very general form.

For the classical nonlinear case, ε_{ij} takes the form, according to Eqs. (1.1.10),

$$\varepsilon_{ij} = \frac{1}{2}(\delta_{ki} u_{k,j} + \delta_{kj} u_{k,i} + u_{k,i} u_{k,j})$$

or

$$\varepsilon_{ij} = \tfrac{1}{2}[\delta_{ki}(e_{kj} + \omega_{kj}) + \delta_{kj}(e_{ki} + \omega_{ki})$$
$$+ (e_{ki} + \omega_{ki})(e_{kj} + \omega_{kj})] \quad (i, j, k = 1, 2, 3). \tag{1.5.9}$$

By substituting this into Eq. (1.5.8), the stress equations of motion and traction boundary conditions are found to be, respectively,

$$[\sigma_{ij}(\delta_{\ell j} + e_{\ell j} + \omega_{\ell j})]_{,i} + f_\ell - \rho \ddot{u}_\ell = 0$$
$$\sigma_{ij}(\delta_{\ell j}(\delta_{\ell j} + \omega_{\ell j}\omega_{\ell j})\nu_i - p_\ell = 0 \qquad (\ell = 1, 2, 3). \tag{1.5.10}$$

These are exactly the same as the results in Eqs. (1.2.4) and (1.2.8).

1.5.2   Pseudo-Variational Equations of Motion

For the first simplified nonlinear case, the linear strains are taken to be negligibly small compared with the rotations, and Eqs. (1.5.9) and (1.5.10) reduce to, respectively,

$$\varepsilon_{ij} = e_{ij} + \tfrac{1}{2}\omega_{ki}\omega_{kj}$$
$$= \tfrac{1}{2}(\delta_{ki} e_{kj} + \delta_{kj} e_{ki} + \omega_{ki}\omega_{kj}) \tag{1.5.11}$$

and

$$[\sigma_{ij}(\delta_{\ell j} + \omega_{\ell j})]_{,i} + f_\ell - \rho \ddot{u}_\ell = 0$$
$$\sigma_{ij}(\delta_{\ell j} + \omega_{\ell j})\nu_i - p_i = 0 \qquad (\ell = 1, 2, 3). \tag{1.5.12}$$

However, Eqs. (1.5.11) and (1.5.12) are not consistent with each other. By substituting Eq. (1.5.11) into Eq. (1.5.8), the stress equations of motion and traction boundary conditions obtained can be shown to be different from the results in Eqs. (1.5.12). On the other hand, it also can be shown that, by starting with Eqs. (1.5.12), a corresponding form of ϵ_{ij} in fact does not exist, and only an expression for $\delta\epsilon_{ij}$ is obtainable. We shall show this through the use of Eq. (1.5.8), according to which the special form of ϵ_{ij}, if existing for Eqs. (1.5.12), must satisfy

$$\int_V \sigma_{ij}\delta\varepsilon_{ij}\,dV = \int_{S_p} \sigma_{ij}(\delta_{\ell j} + \omega_{\ell j})\nu_i\,\delta u_\ell\,dS$$
$$- \int_V [\sigma_{ij}(\delta_{\ell j} + \omega_{\ell j})]_{,i}\,\delta u_\ell\,dV.$$

By transforming the surface integral into a volume integral, this becomes

$$\int_V \sigma_{ij}\delta\varepsilon_{ij}\,dV = \int_V \sigma_{ij}(\delta_{\ell j} + \omega_{\ell j})\delta u_{\ell,i}\,dV$$
$$= \int_V \sigma_{ij}[\delta(e_{ij} + \omega_{ji} + \tfrac{1}{2}\omega_{ki}\omega_{kj}) + \omega_{kj}\delta e_{ki}]\,dV,$$

from which, since $\omega_{ji} = -\omega_{ij}$,

$$\delta\varepsilon_{ij} = \delta(e_{ij} + \tfrac{1}{2}\omega_{ki}\omega_{kj}) + \omega_{kj}\delta e_{ki}. \tag{1.5.13}$$

An expression of ϵ_{ij} thus is not obtainable, although $\delta\epsilon_{ij}$ has been determined. The converse situation has been noted by Biot (1939); namely, if $\delta\epsilon_{ij}$ as given by Eq. (1.5.13) is accepted, then the stress equations of motion and traction boundary conditions in Eqs. (1.5.12) can be derived.

It is clear that the variational principle and variational equation correspond to each other exactly only in the classical nonlinear case, but not in the two simplified nonlinear cases. The variational equations of motion given by Eqs. (1.4.7) and (1.4.8) for the simplified cases must be considered as simplified versions of Eq. (1.4.5) or (1.4.6) for the classical case. Since the former cannot be derived exactly from a variational principle, they are called pseudo-variational equations of motion (Yu 1991, 1995a). Of the two pseudo-variational equations, Eq. (1.4.8) for the new simplified case has been found to be particularly useful in the development of dynamical modeling for large deflections of beams, plates, and shells. This will be demonstrated in later chapters in this book.

1.6   Generalized Hamilton's Principle and Variational Equation of Motion

The principle of virtual work and Hamilton's principle are variational principles for displacements. In contrast, the well-known Castigliano's theorem of least work is a variational principle for stresses. Reissner (1950) presented a variational principle for both displacements and stresses. Hu (1954) and Washizu (1955, 1968) further considered variations of strains as well as variations of displacements and stresses in the formulation of a variational principle in elastostatics. Yu (1964) extended the works of Hu and Washizu to the dynamic case in nonlinear elasticity theory, and the results were a generalized Hamilton's principle and the associated generalized variational equation of motion. These results are presented here.

The generalized Hamilton's principle has the form

$$\delta\int_{t_0}^{t_1} L\,dt = \delta\int_{t_0}^{t_1} (T - U + W)\,dt = 0, \tag{1.6.1}$$

where $L = T - U + W$ is the generalized Lagrangian function with

$$T = \int_V \tfrac{1}{2}\rho\dot{u}_i\dot{u}_i\,dV$$

$$U = \int_V [\sigma_{ij}(\varepsilon_{ij} - \epsilon_{ij}) + U_0]\,dV \quad (i, j = 1, 2, 3) \tag{1.6.2}$$

$$W = \int_V f_i u_i\,dV + \int_{S_p} \bar{p}_i u_i\,dS + \int_{S_u} p_i(u_i - \bar{u}_i)\,dS.$$

Equation (1.6.1) has the same form as in the ordinary Hamilton's principle, but U and W are now different. In Eqs. (1.6.2), Cartesian tensor notations and summation convention for repeated indices again are adopted. Most of the notations are similar to those used in previous sections. Among the new notations, ε_{ij} are expressions of ϵ_{ij} as functions of the derivatives of u_i, an overdot denotes differentiation with respect to time, and an overbar denotes the prescribed value of a quantity.

The variations of the displacements, strains, and stresses are taken independently. We thus have

$$\delta \int_{t_1}^{t_0} T\, dt = \int_V \rho \dot{u}_i \delta u_i\, dV \Big|_{t_0}^{t_1} - \int_{t_0}^{t_1} dt \int_V \rho \ddot{u}_i \delta u_i\, dV \tag{1.6.3}$$

$$\delta \int_{t_0}^{t_1} U\, dt = \int_{t_0}^{t_1} dt \int_V \left[\sigma_{ij} \delta \varepsilon_{ij} + (\varepsilon_{ij} - \epsilon_{ij}) \delta \sigma_{ij} \right.$$
$$\left. - \left(\sigma_{ij} - \frac{\partial U_0}{\partial \epsilon_{ij}} \right) \delta \epsilon_{ij} \right] dV \tag{1.6.4}$$

$$\delta \int_{t_0}^{t_1} W\, dt = \int_{t_0}^{t_1} \left[\int_V f_i \delta u_i\, dV + \int_{S_p} \bar{p}_i \delta u_i\, dS + \int_{S_u} (u_i - \bar{u}_i) \delta p_i\, dS \right] dt. \tag{1.6.5}$$

Since T is the same as in the ordinary Hamilton principle, δT is still the same as before. Also, since ε_{ij} are functions of the first derivatives of displacements as before, we again have

$$\int_V \sigma_{ij} \delta \epsilon_{ij}\, dV = \int_{S_p} \frac{1}{2} \left[\sigma_{ij} \left(\frac{\partial \varepsilon_{ij}}{\partial e_{mn}} + \frac{\partial \varepsilon_{ij}}{\partial \omega_{mn}} \right) v_n \delta_{\ell m} + \cdots, \tag{1.6.6} \right.$$

which is the same as Eq. (1.5.7). By virtue of Eqs. (1.6.3) through (1.6.6), Eq. (1.6.1) yields

$$\int_{t_0}^{t_1} dt \int_V \left[\frac{1}{2} \left\{ \sigma_{ij} \left(\frac{\partial \varepsilon_{ij}}{\partial e_{mn}} + \frac{\partial \varepsilon_{ij}}{\omega_{mn}} \right) \right\}_{,n} \delta_{\ell m} \right.$$
$$\left. + \frac{1}{2} \left\{ \sigma_{ij} \left(\frac{\partial \varepsilon_{ij}}{\partial e_{mn}} - \frac{\partial \varepsilon_{ij}}{\partial \omega_{mn}} \right) \right\}_{,m} \delta_{\ell n} + f_\ell - \rho \ddot{u}_\ell \right] \delta u_\ell\, dV$$
$$- \int_{t_0}^{t_1} dt \int_{S_p} \left[\frac{1}{2} \sigma_{ij} \left(\frac{\partial \varepsilon_{ij}}{\partial e_{mn}} + \frac{\partial \varepsilon_{ij}}{\partial \omega_{mn}} \right) v_n \delta_{\ell m} \right.$$
$$\left. + \frac{1}{2} \sigma_{ij} \left(\frac{\partial \varepsilon_{ij}}{\partial e_{mn}} - \frac{\partial \varepsilon_{ij}}{\partial \omega_{mn}} \right) v_m \delta_{\ell n} - \bar{p}_\ell \right] \delta u_\ell\, dS$$
$$+ \int_{t_0}^{t_1} dt \int_V \left(\sigma_{ij} - \frac{\partial U_0}{\partial \epsilon_{ij}} \right) \delta \epsilon_{ij}\, dV$$
$$- \int_{t_0}^{t_1} dt \int_V (\varepsilon_{ij} - \epsilon_{ij}) \delta \sigma_{ij}\, dV$$

$$+ \int_{t_0}^{t_1} dt \int_{S_u} (u_i - \bar{u}_i)\delta p_i \, dS \, dt = 0. \tag{1.6.7}$$

Since the variations δu_i, $\delta \epsilon_{ij}$, and $\delta \sigma_{ij}$ are arbitrary throughout V, δu_i is arbitrary on S_p, and δp_i is arbitrary on S_u, their coefficients in the five integrands in Eq. (1.6.7) must vanish independently. This yields in succession the stress equations of motion, traction boundary conditions, stress–strain relations, strain–displacement relations, and displacement boundary conditions. These constitute the complete system of equations for large elastic deformations. According to Eqs. (1.6.1), the generalized Hamilton's principle thus may be stated as follows:

The displacements, strains (defined in the manner of Green), and stresses (defined in the manner of Kirchhoff) which, over the time interval from t_0 to t_1, satisfy the stress equations of motion and the stress–strain–displacement relations throughout V, the traction boundary conditions over S_p, and the displacement boundary conditions over S_u, are determined by the vanishing of the variation of the time integral of the generalized Lagrangian function over that time interval, provided that the variations of the displacements, strains, and stresses be taken independently and simultaneously, that the variations of the displacements vanish at t_0 and t_1 throughout the body, and that the variations of the displacements and tractions be in compliance with the prescribed boundary conditions.

Equation (1.6.7) is the generalized variational equation of motion. The generalized Hamilton's principle and variational equation of motion are naturally applicable to small deformations as a special case. If the variations are restricted to those of displacements only, the results reduce to the ordinary Hamilton's principle and variational equation of motion discussed in Section 1.4.

1.7   Stress–Strain Relations in Nonlinear Elasticity

Now that Kirchhoff's stress tensor and Green's nonlinear strain tensor have been adopted, the stress-strain relations in nonlinear elasticity can be written in a form similar to, as well as reducible to, those in linear elasticity. This is based on the assumed existence of the strain energy function as introduced in Eqs. (1.3.6). As mentioned earlier, the components of Green's nonlinear strain tensor are direct measures of the extensional strains and shear angles. Specifically, the extensional strains and shear angles disappear when the nonlinear strain components vanish. Being components of a tensor, the nonlinear and the associated linear shearing strains are written with a factor $\frac{1}{2}$, and these will be referred to as the tensorial shearing strains. As mentioned earlier, they are different from the ordinary engineering shearing strains, which do not include the factor $\frac{1}{2}$. The engineering shearing strains are therefore equal to twice the corresponding tensorial shearing strains in both linear and nonlinear cases.

1.7.1 Generalized Hooke's Law

Hooke's law in its original form is associated with simple extension of a one-dimensional bar made of an isotropic, elastic material. The generalized Hooke's law extends this to the three-dimensional state of stress in an anisotropic material. We shall find it convenient to use the following contracted notations:

$$\sigma_1 = \sigma_{xx}, \quad \sigma_2 = \sigma_{yy}, \quad \sigma_3 = \sigma_{zz}, \quad \sigma_4 = \sigma_{yz}, \quad \sigma_5 = \sigma_{zx}, \quad \sigma_6 = \sigma_{xy},$$
$$\epsilon_1 = \epsilon_{xx}, \quad \epsilon_2 = \epsilon_{yy}, \quad \epsilon_3 = \epsilon_{zz}, \quad \epsilon_4 = 2\epsilon_{yz}, \quad \epsilon_5 = 2\epsilon_{zx}, \quad \epsilon_6 = 2\epsilon_{xy}. \tag{1.7.1}$$

In particular, ϵ_4, ϵ_5, and ϵ_6 are the engineering shearing strains and ϵ_{yz}, ϵ_{zx}, and ϵ_{xy} are the corresponding tensorial shearing strains. The generalized Hooke's law may now be written in the following matrix form:

$$
\begin{bmatrix} \sigma_1 \\ \sigma_2 \\ \sigma_3 \\ \sigma_4 \\ \sigma_5 \\ \sigma_6 \end{bmatrix}
=
\begin{bmatrix}
c_{11} & c_{12} & c_{13} & c_{14} & c_{15} & c_{16} \\
c_{21} & c_{22} & c_{23} & c_{24} & c_{25} & c_{26} \\
c_{31} & c_{32} & c_{33} & c_{34} & c_{35} & c_{36} \\
c_{41} & c_{42} & c_{43} & c_{44} & c_{45} & c_{46} \\
c_{51} & c_{52} & c_{53} & c_{54} & c_{55} & c_{56} \\
c_{61} & c_{62} & c_{63} & c_{64} & c_{65} & c_{66}
\end{bmatrix}
\begin{bmatrix} \epsilon_1 \\ \epsilon_2 \\ \epsilon_3 \\ \epsilon_4 \\ \epsilon_5 \\ \epsilon_6 \end{bmatrix},
\tag{1.7.2}
$$

where $c_{11}, \ldots,$ are the stiffnesses. Equation (1.7.2) also may be written in Cartesian tensor notation as

$$\sigma_i = c_{ij}\,\epsilon_j \quad (i, j = 1, 2, \ldots, 6). \tag{1.7.3}$$

By inversion, this becomes

$$\epsilon_i = s_{ij}\,\sigma_j \quad (i, j = 1, 2, \ldots, 6), \tag{1.7.4}$$

where s_{ij} are the compliances. The maximum number of independent stiffnesses or compliances is 36, but this reduces to 21 because of symmetry of the stress and strain tensors, namely,

$$c_{ij} = c_{ji}, \quad s_{ij} = s_{ji}.$$

In the general case, the strain energy function has the form

$$U_0 = \tfrac{1}{2} c_{ij}\epsilon_i\epsilon_j \quad (i, j = 1, \ldots, 6),$$

from which

$$\frac{\partial U_0}{\partial \epsilon_i} = \sigma_i \quad (i = 1, 2, \ldots, 6).$$

These are the same as Eqs. (1.3.6) and confirm the existence of U_0.

Consider a material that has a *plane of elastic symmetry*, say, the xy-plane. This means that the stiffnesses c_{ij} are invariant under a coordinate transformation involving the reversal of the z-axis. We then have

$$c_{14} = c_{15} = c_{24} = c_{25} = c_{34} = c_{35} = c_{46} = c_{56} = 0,$$

and the stiffness matrix in Eq. (1.7.2) reduces to the form

$$[c_{ij}] = \begin{bmatrix} c_{11} & c_{12} & c_{13} & 0 & 0 & c_{16} \\ c_{21} & c_{22} & c_{23} & 0 & 0 & c_{26} \\ c_{31} & c_{32} & c_{33} & 0 & 0 & c_{36} \\ 0 & 0 & 0 & c_{44} & c_{45} & 0 \\ 0 & 0 & 0 & c_{54} & c_{55} & 0 \\ c_{61} & c_{62} & c_{63} & 0 & 0 & c_{66} \end{bmatrix}. \tag{1.7.5}$$

The number of independent stiffnesses is reduced to 13.

If a material has elastic symmetry with respect to two mutually orthogonal planes, it also will have elastic symmetry with respect to a third plane that is orthogonal to the other two. In this case, the material is said to be *orthotropic*, for which we have

$$c_{16} = c_{26} = c_{36} = c_{45} = 0,$$

and Eq. (1.7.5) becomes

$$[c_{ij}] = \begin{bmatrix} c_{11} & c_{12} & c_{13} & 0 & 0 & 0 \\ c_{21} & c_{22} & c_{23} & 0 & 0 & 0 \\ c_{31} & c_{32} & c_{33} & 0 & 0 & 0 \\ 0 & 0 & 0 & c_{44} & 0 & 0 \\ 0 & 0 & 0 & 0 & c_{55} & 0 \\ 0 & 0 & 0 & 0 & 0 & c_{66} \end{bmatrix}. \tag{1.7.6}$$

There are therefore nine independent stiffnesses.

If one of the three coordinate planes, say, the yz-plane, is isotropic in the sense that the material properties in that plane are independent of direction, then

$$c_{33} = c_{22}, \quad c_{13} = c_{12}, \quad c_{55} = c_{66}, \quad c_{44} = \frac{c_{22} - c_{23}}{2},$$

and Eq. (1.7.6) reduces to

$$[c_{ij}] = \begin{bmatrix} c_{11} & c_{12} & c_{12} & 0 & 0 & 0 \\ c_{12} & c_{22} & c_{23} & 0 & 0 & 0 \\ c_{12} & c_{23} & c_{22} & 0 & 0 & 0 \\ 0 & 0 & 0 & \frac{c_{22}-c_{23}}{2} & 0 & 0 \\ 0 & 0 & 0 & 0 & c_{66} & 0 \\ 0 & 0 & 0 & 0 & 0 & c_{66} \end{bmatrix}. \tag{1.7.7}$$

There are five independent elastic constants, and the material is said to be *transversely isotropic*.

Finally, for complete *isotropy*, we have further

$$c_{22} = c_{11}, \quad c_{23} = c_{12}, \quad c_{66} = \frac{c_{11} - c_{22}}{2},$$

and Eq. (1.7.7) takes the form

$$[c_{ij}] = \begin{bmatrix} c_{11} & c_{12} & c_{12} & 0 & 0 & 0 \\ c_{12} & c_{11} & c_{12} & 0 & 0 & 0 \\ c_{12} & c_{12} & c_{11} & 0 & 0 & 0 \\ 0 & 0 & 0 & \frac{c_{11}-c_{12}}{2} & 0 & 0 \\ 0 & 0 & 0 & 0 & \frac{c_{11}-c_{12}}{2} & 0 \\ 0 & 0 & 0 & 0 & 0 & \frac{c_{11}-c_{12}}{2} \end{bmatrix}. \tag{1.7.8}$$

The stress–strain relations are thus

$$\sigma_{xx} = c_{11}\epsilon_{xx} + c_{12}(\epsilon_{yy} + \epsilon_{zz}), \ldots$$
$$\sigma_{yz} = (c_{11} - c_{12})\epsilon_{yz}, \ldots.$$

These often are written in the form

$$\sigma_{xx} = (\lambda + 2\mu)\epsilon_{xx} + \lambda(\epsilon_{yy} + \epsilon_{zz}), \ldots$$
$$\sigma_{yz} = 2\mu\epsilon_{yz}, \ldots, \tag{1.7.9}$$

where only two independent elastic constants remain; these are the Lamé constants

$$\lambda = c_{12}, \quad \mu = \frac{c_{11} - c_{12}}{2}.$$

1.7.2 Engineering Constants

Let us examine the relations between the stiffnesses and compliances, together with the commonly used engineering constants, which include Young's modulus, Poisson's ratio, and the shear modulus. We start with the strain–stress relations in Eqs. (1.7.4) for an orthotropic material in the following matrix form:

$$\begin{bmatrix} \epsilon_1 \\ \epsilon_2 \\ \epsilon_3 \\ \epsilon_4 \\ \epsilon_5 \\ \epsilon_6 \end{bmatrix} = \begin{bmatrix} s_{11} & s_{12} & s_{13} & 0 & 0 & 0 \\ s_{21} & s_{22} & s_{23} & 0 & 0 & 0 \\ s_{31} & s_{32} & s_{33} & 0 & 0 & 0 \\ 0 & 0 & 0 & s_{44} & 0 & 0 \\ 0 & 0 & 0 & 0 & s_{55} & 0 \\ 0 & 0 & 0 & 0 & 0 & s_{66} \end{bmatrix} \begin{bmatrix} \sigma_1 \\ \sigma_2 \\ \sigma_3 \\ \sigma_4 \\ \sigma_5 \\ \sigma_6 \end{bmatrix}. \tag{1.7.10}$$

The compliances are related to the engineering constants by

$$s_{11} = \frac{1}{E_{11}}, \quad s_{12} = \frac{-\nu_{12}}{E_{11}}, \quad s_{13} = \frac{-\nu_{13}}{E_{11}}$$
$$s_{22} = \frac{1}{E_{22}}, \quad s_{23} = \frac{-\nu_{23}}{E_{22}}, \quad s_{33} = \frac{1}{E_{33}} \tag{1.7.11}$$
$$s_{44} = \frac{1}{2G_{23}}, \quad s_{55} = \frac{1}{2G_{13}}, \quad s_{66} = \frac{1}{2G_{12}},$$

where E_{ij} is Young's modulus in the ith-direction, v_{ij} is Poisson's ratio reflecting the contraction in the jth-direction due to a tension in the ith-direction, and G_{ij} is the shear modulus with respect to the i, j-directions. Of the twelve engineering constants in Eq. (1.7.10), only nine are independent. If these are chosen to be E_{11}, E_{22}, E_{33}, v_{12}, v_{13}, v_{23}, G_{12}, G_{13}, and G_{23}, then the other three, v_{21} v_{31}, and v_{32}, are determined by

$$E_{11}v_{21} = E_{22}v_{12}, \quad E_{22}v_{32} = E_{33}v_{23}, \quad E_{33}v_{13} = E_{11}v_{31}.$$

For a transversely isotropic material for which the plane $x_1 = 0$ is the plane of isotropy, we have

$$E_{33} = E_{22}, \quad G_{13} = G_{12}, \quad v_{13} = v_{12}, \quad G_{23} = \frac{E_{22}}{2(1 + v_{23})},$$

and there are only five independent engineering constants. For a completely isotropic material, these finally reduce to two, and we have

$$E_{11} = E_{22} = E_{33} = E$$

$$G_{12} = G_{23} = G_{31} = G = \frac{E}{2(1 + v)}. \tag{1.7.12}$$

The stiffnesses c_{ij} can be solved in terms of the engineering constants by substituting Eqs. (1.7.11) into (1.7.10) and inverting the result.

1.7.3 Plane Stress

The state of plane stress is often assumed. If we choose $\sigma_{zz} = \sigma_{zy} = \sigma_{zx} = 0$ or, equivalently, $\sigma_3 = \sigma_4 = \sigma_5 = 0$, then Eqs. (1.7.10) reduce to, with compliances substituted from Eqs. (1.7.11),

$$\begin{bmatrix} \epsilon_1 \\ \epsilon_2 \\ \epsilon_3 \end{bmatrix} = \begin{bmatrix} \frac{1}{E_{11}} & \frac{-v_{12}}{E_{11}} & 0 \\ \frac{-v_{12}}{E_{11}} & \frac{1}{E_{22}} & 0 \\ 0 & 0 & \frac{1}{2G_{12}} \end{bmatrix} \begin{bmatrix} \sigma_1 \\ \sigma_2 \\ \sigma_3 \end{bmatrix}. \tag{1.7.13}$$

By inversion, we find

$$\begin{bmatrix} \sigma_1 \\ \sigma_2 \\ \sigma_3 \end{bmatrix} = \begin{bmatrix} Q_{11} & Q_{12} & 0 \\ Q_{12} & Q_{22} & 0 \\ 0 & 0 & Q_{66} \end{bmatrix} \begin{bmatrix} \epsilon_1 \\ \epsilon_2 \\ \epsilon_3 \end{bmatrix}, \tag{1.7.14}$$

where

$$Q_{11} = \frac{E_{11}}{1 - v_{12}E_{22}/E_{11}}$$

$$Q_{12} = \frac{v_{12}E_{22}}{1 - v_{12}E_{22}/E_{11}}$$

$$Q_{22} = \frac{E_{22}}{1 - v_{12}E_{22}/E_{11}}$$

$$Q_{66} = 2G_{12}$$

are the reduced stiffnesses for plane stress. For an isotropic material, Eqs. (1.7.12) are again valid.

References

Biot, M.A. (1939) Nonlinear Theory of Elasticity and the Linearized Case for a Body under Initial Stress. *Philosophical Magazine*, Vol. 27, Ser. 7, pp. 468–489.

Fung, Y.C. (1965) *Foundations of Solid Mechanics.* Prentice-Hall, Englewood Cliffs, New Jersey.

Hu, H.C. (1954) On Some Variational Principles in the Theory of Elasticity and the Theory of Plasticity. *Acta Physica Sinica*, Vol. 10, p. 259. (Also, *Scientia Sinica*, Vol. 4, p. 33, 1955.)

Kirchhoff, G. (1883) *Vorlesungen über math. Physik, Mechanik*, 3rd Ed. Leipzig.

Love, A.E.H. (1927) *A Treatise on the Mathematical Theory of Elasticity*, 4th Ed. Cambridge University Press. (Also, Dover Publications, first American edition, 1944.)

Marguerre, K. (1962) *Handbook of Engineering Mechanics*, edited by W. Flügge, Chap. 33. McGraw-Hill, New York.

Novozhilov, V.V. (1948) *Foundations of the Nonlinear Theory of Elasticity*. First Russian edition: Gostekhizdat; English translation: Graylock Press, Rochester, New York, 1953.

Reissner, E. (1950) On a Variational Theorem in Elasticity. *Journal of Mathematics and Physics*, Vol. 29, pp. 90–95.

Timoshenko, S. and J.N. Goodier. (1970) *Theory of Elasticity*, 3rd Ed. McGraw-Hill, New York.

Washizu, K. (1955) *On the Variational Principles of Elasticity and Plasticity*. M.I.T. Aeroelastic Structures Research Laboratory Technical Report 25.18, Cambridge, Massachusetts.

Washizu, K. (1968) *Variational Methods in Elasticity and Plasticity*. Pergamon Press, New York.

Yu, Y.Y. (1964) Generalized Hamilton's Principle and Variational Equation of Motion in Nonlinear Elasticity Theory, with Application to Plate Theory. *Journal of the Acoustical Society of America*, Vol. 36, pp. 111–120.

Yu, Y.Y. (1991) On Equations for Large Deflections of Elastic Plates and Shallow Shells. *Mechanics Research Communications*, Vol. 18, pp. 373–384.

Yu, Y.Y. (1995a) On the Ordinary, Generalized, and Pseudo-Variational Equations of Motion in Nonlinear Elasticity, Piezoelectricity, and Classical Plate Theory. *Journal of Applied Mechanics*, Vol. 62, pp. 471–478.

Yu, Y.Y. (1995b) Some Recent Advances in Linear and Nonlinear Dynamical Modeling of Elastic and Piezoelectric Plates. *Journal of Intelligent Material Systems and Structures*, Vol. 6, pp. 237–254.

2

Linear Vibrations of Plates Based on Elasticity Theory

In this chapter, we discuss some exact solutions for linear vibrations of plates derived from the exact elasticity theory. In later chapters, approximate equations for plates are first derived from elasticity theory, and solutions are then in turn obtained from the plate equations. The plate equations are considered to be approximate from the standpoint of the elasticity theory, and solutions obtained from such equations are thus also approximate. Generally speaking, exact solutions based on elasticity are difficult to find, and most of them deal with plates that extend to infinity. In fact, no solutions in closed form are known to exist for vibrations of plates with unrestricted dimensional ratios and with traction-free boundary surfaces. On the other hand, we are fortunate to find the small number of exact solutions for plates that are available. Among other things, these do provide an overall perspective and insight by making it possible for us to evaluate the accuracy and limitations of the approximate equations and solutions of plates.

All exact solutions discussed in this chapter are based on linear elasticity theory. We start with the famous solution of Rayleigh (1888) and Lamb (1889) for the vibration of an infinite isotropic plate with free faces. Although their frequency equation has a very simple appearance, it was not until 70 years later that Mindlin (1960) explored the full implications and proceeded toward the solution of problems involving high frequencies in plates with boundaries. A rich source of information on related subjects may be found in the two volumes of Mindlin's collected papers (Deresiewicz *et al.* 1989). By extending the treatment for a single-layered plate, we obtained an exact solution for the vibration of an infinite three-layered sandwich plate (Yu 1959a,b, 1960, 1962, 1995), which is also discussed in detail in this chapter.

2.1   Equations of Linear Elasticity Theory

In the absence of body forces, the stress equations of motion in linear elasticity theory are, as can be written from the linear variational equation (1.4.9),

$$\frac{\partial \sigma_{xx}}{\partial x} + \frac{\partial \sigma_{yx}}{\partial y} + \frac{\partial \sigma_{zx}}{\partial z} = \rho \ddot{u}_x$$

$$\frac{\partial \sigma_{xy}}{\partial x} + \frac{\partial \sigma_{yy}}{\partial y} + \frac{\partial \sigma_{zy}}{\partial z} = \rho \ddot{u}_y \qquad (2.1.1)$$

$$\frac{\partial \sigma_{xz}}{\partial x} + \frac{\partial \sigma_{yz}}{\partial y} + \frac{\partial \sigma_{zz}}{\partial z} = \rho \ddot{u}_z.$$

For an isotropic elastic solid, the nonlinear stress–strain relations were given by Eqs. (1.7.9), which yield for the linear case,

$$\sigma_{xx} = (\lambda + 2\mu)e_{xx} + \lambda(e_{yy} + e_{zz})$$

$$= (\lambda + 2\mu)\left(\frac{\partial u_x}{\partial x}\right) + \lambda\left(\frac{\partial u_y}{\partial y} + \frac{\partial u_z}{\partial z}\right), \ldots \qquad (2.1.2)$$

$$\sigma_{yz} = 2\mu e_{yz} = \mu\left(\frac{\partial u_x}{\partial y} + \frac{\partial u_y}{\partial x}\right), \ldots$$

Substitution of Eqs. (2.1.2) into (2.1.1) yields the following displacement equations of motion:

$$\mu\nabla^2 u_x + (\lambda + \mu)\frac{\partial \Delta}{\partial x} = \rho \frac{\partial^2 u_x}{\partial t^2}$$

$$\mu\nabla^2 u_y + (\lambda + \mu)\frac{\partial \Delta}{\partial y} = \rho \frac{\partial^2 u_y}{\partial t^2} \qquad (2.1.3)$$

$$\mu\nabla^2 u_z + (\lambda + \mu)\frac{\partial \Delta}{\partial z} = \rho \frac{\partial^2 u_z}{\partial t^2},$$

where the Laplace operator and dilatation are, respectively,

$$\nabla^2 = \frac{\partial^2}{\partial x^2} + \frac{\partial^2}{\partial y^2} + \frac{\partial^2}{\partial z^2}, \quad \Delta = \frac{\partial u_x}{\partial x} + \frac{\partial u_y}{\partial y} + \frac{\partial u_z}{\partial z}.$$

The displacements may be expressed further in terms of four potential functions as follows:

$$u_x = \frac{\partial \phi}{\partial x} + \frac{\partial H_3}{\partial y} - \frac{\partial H_2}{\partial z}$$

$$u_y = \frac{\partial \phi}{\partial y} + \frac{\partial H_1}{\partial z} - \frac{\partial H_3}{\partial x} \qquad (2.1.4)$$

$$u_z = \frac{\partial \phi}{\partial z} + \frac{\partial H_2}{\partial x} - \frac{\partial H_1}{\partial y}$$

provided that

$$\frac{\partial H_1}{\partial x} + \frac{\partial H_2}{\partial y} + \frac{\partial H_3}{\partial z} = 0. \tag{2.1.5}$$

It follows from these that

$$\nabla^2 \phi = \Delta, \quad \nabla^2 H_1 = -2\omega_{yz}, \quad \nabla^2 H_2 = -2\omega_{zx}, \quad \nabla^2 H_3 = -2\omega_{xy}, \tag{2.1.6}$$

where the rotations are as defined in Eqs. (1.1.7). Thus, ϕ gives rise to dilatation and H_1, H_2, and H_3 give rise to rotations.

The displacement equations of motion are satisfied if the four potentials satisfy the wave equations

$$c_\alpha^2 \nabla^2 \phi = \frac{\partial^2 \phi}{\partial t^2}, \quad c_\beta^2 \nabla^2 H_i = \frac{\partial^2 H_i}{\partial t^2} \quad (i = 1, 2, 3), \tag{2.1.7}$$

where

$$c_\alpha = \sqrt{\frac{\lambda + 2\mu}{\rho}}, \quad c_\beta = \sqrt{\frac{\mu}{\rho}}. \tag{2.1.8}$$

are the velocities of propagation of dilatational and rotational plane waves, respectively, in an infinite elastic solid. The displacements associated with these waves are parallel and at right angles, respectively, to the direction of wave propagation; the waves that are at right angles to the direction of propagation can take place in either a vertical or horizontal plane. In seismology, these are often called the P, SV, and SH waves, respectively. While the two types of waves, dilatational and rotational, can exist independently in an infinite solid, they generally become coupled with each other in a plate if any part of the plate boundary is free of traction. This is the primary source of complexity of the theory of vibrations of finite plates with boundary surfaces, as pointed out by Mindlin (1960).

2.2 Rayleigh–Lamb Solution for Plane-Strain Modes of Vibration in an Infinite Plate

The exact solution to the problem of vibration of an infinite plate with traction-free boundary planes was due to Rayleigh (1888) and Lamb (1889). We choose the boundary planes of the plate at $z = \pm h$ and, for plane strain, let the displacements take the form

$$u_x = u_x(x, z, t), \quad u_y = 0, \quad u_z = u_z(x, z, t),$$

according to which Eqs. (2.1.4) reduce to

$$u_x = \frac{\partial \phi}{\partial x} - \frac{\partial H_2}{\partial z}, \quad u_z = \frac{\partial \phi}{\partial z} + \frac{\partial H_2}{\partial x}. \tag{2.2.1}$$

The two potentials ϕ and H_2 still satisfy the wave equations (2.1.7), but the Laplace operator now has the simpler two-dimensional form

$$\nabla^2 = \frac{\partial^2}{\partial x^2} + \frac{\partial^2}{\partial z^2}.$$

These potentials have the solutions

$$\phi = f(z)\sin\xi x\, e^{i\omega t}$$
$$H_2 = -g(z)\cos\xi x\, e^{i\omega t}, \tag{2.2.2}$$

where ω is the frequency and ξ is the wave number along the x-direction, related to the wavelength ℓ by $\xi = 2\pi/\ell$. The functions f and g satisfy the equations

$$f'' + \alpha^2 f = 0, \quad g'' + \beta^2 g = 0, \tag{2.2.3}$$

where a prime denotes differentiation with respect to z, and

$$\alpha^2 = \frac{\omega^2}{c_\alpha^2} - \xi^2, \quad \beta^2 = \frac{\omega^2}{c_\beta^2} - \xi^2. \tag{2.2.4}$$

The solutions of Eqs. (2.2.3) are simply

$$f = A\sin\alpha z + B\cos\alpha z$$
$$g = A\sin\beta z + B\cos\beta z, \tag{2.2.5}$$

and α and β thus are wave numbers across the thickness of the plate.

The displacements are finally obtained from Eqs. (2.2.1), (2.2.2), and (2.2.5), and the stresses in turn from Eqs. (2.1.2). The nonzero displacement and stress components are

$$
\begin{aligned}
u_x &= (\xi f + g')\cos\xi x\, e^{i\omega t}\\
u_z &= (f' + \xi g)\sin\xi x\, e^{i\omega t}\\
\sigma_{xx} &= -\mu[(\beta^2 + \xi^2 - 2\alpha^2)f + 2\xi g']\sin\xi x\, e^{i\omega t}\\
\sigma_{yy} &= -\lambda(\alpha^2 + \xi^2)f\sin\xi x\, e^{i\omega t}\\
\sigma_{zz} &= -\mu[(\beta^2 - \xi^2)f - 2\xi g']\sin\xi x\, e^{i\omega t}\\
\sigma_{zx} &= \mu[2\xi f' + (\xi^2 - \beta^2)g]\cos\xi x\, e^{i\omega t}.
\end{aligned}
\tag{2.2.6}
$$

For an infinite plate with traction-free boundary planes, the boundary conditions are

$$\sigma_{zz} = \sigma_{zx} = 0 \quad \text{at} \quad z = \pm h.$$

Substitution of the stresses from Eqs. (2.2.6) into the boundary conditions yields the symmetric and antisymmetric modes of vibration, uncoupled from each other. The mode shape and frequency equation of symmetric modes are given by

$$\frac{B}{C} = -\frac{2\xi\beta \cos \beta h}{(\xi^2 - \beta^2) \cos \alpha h}$$

$$\frac{\tan \beta h}{\tan \alpha h} = -\frac{4\xi^2 \alpha \beta}{(\xi^2 - \beta^2)^2}, \qquad (2.2.7)$$

and those of antisymmetric modes are given by

$$\frac{A}{D} = \frac{2\xi\beta \sin \beta h}{(\xi^2 - \beta^2) \sin \alpha h}$$

$$\frac{\tan \beta h}{\tan \alpha h} = -\frac{(\xi^2 - \beta^2)^2}{4\xi^2 \alpha \beta}. \qquad (2.2.8)$$

The frequency equations (2.2.7) and (2.2.8) obtained by Rayleigh and Lamb are deceptively simple in appearance and have been the subject of extensive studies by many authors. As was mentioned earlier, it was not until 70 years later that the full implications of these equations became understood thoroughly enough, mostly due to the effort of Mindlin (1960). Rather than covering the full details, we shall only discuss those aspects of the Rayleigh–Lamb solution that will be particularly germane to our studies in this book.

2.2.1   Frequency Spectrum

Results obtained from the frequency equations (2.2.7) and (2.2.8) as given by Mindlin (1960) are reproduced in Figure 2.2.1. The branches of the frequency spectrum of an infinite plate are shown in dashed and full lines for the antisymmetric and symmetric modes, respectively. The abscissa gives the real and imaginary parts x and y of the complex variable z, as related to the wave number ξ by

$$z = x + iy = \frac{2\xi h}{\pi}.$$

The ordinate is the frequency ratio

$$\Omega = \frac{\omega}{\omega_s},$$

where

$$\omega_s = \frac{\pi c_\beta}{2h}$$

is the frequency of the lowest simple thickness-shear mode of vibration of the infinite plate, which will be discussed in the next section. While the frequency must be real and positive, the wave numbers ξ, α, and β may be real, imaginary, or

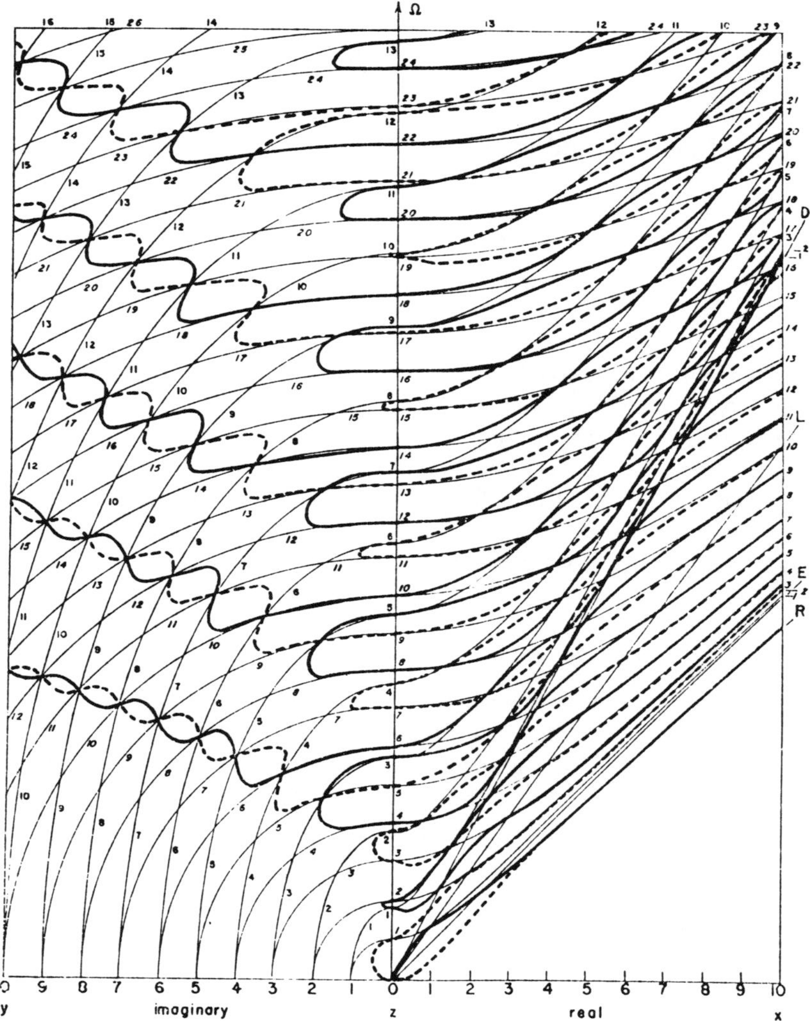

Fig. 2.2.1. Frequency spectrum of an infinite plate for real and imaginary wave numbers and $\nu = 0.31$ (after Mindlin).

complex. In rectangular coordinates, real wave numbers correspond to mode shapes that are described by trigonometric functions; imaginary wave numbers correspond to modes described by hyperbolic or real exponential functions; and complex wave numbers correspond to modes described by the products of a combination of these two types of functions.

To appreciate the wide frequency range covered in Figure 2.2.1, consider the example of a plate made of steel, for which the rotational or shear wave velocity is

$$c_\beta = 3220 \text{ meters/second.}$$

It then is easily calculated that, for a plate thickness of 1 cm, the lowest simple thickness-shear frequency ω_s is slightly over 160,000 hertz and the total frequency range covered by Figure 2.2.1 is well over 4,000,000 hertz. While these frequencies may appear to be extraordinarily high for ordinary engineering applications, some of Mindlin's studies were to fill the needs to learn about high-frequency vibrations of crystal plates used as resonators.

For a more limited frequency range, we show in Figure 2.2.2 the enlarged views of several of the lowest branches of the frequency spectrum for a limited range of the real wave number only and for Poisson's ratio equal to 1/4, 1/3, and 1/2.5 (Mindlin 1955). Again, full lines are for symmetric modes and dashed lines are for antisymmetric modes. The ordinate is still the same frequency ratio Ω as in Figure 2.2.1, and the abscissa is now ξh.

In Figures 2.2.1 and 2.2.2, frequencies have been plotted versus wave numbers in a dimensionless form. Phase and group velocities may readily be determined from these plots. The slope of a straight line connecting a point on a branch with the origin gives the phase velocity, and the slope of the branch at the point gives the group velocity.

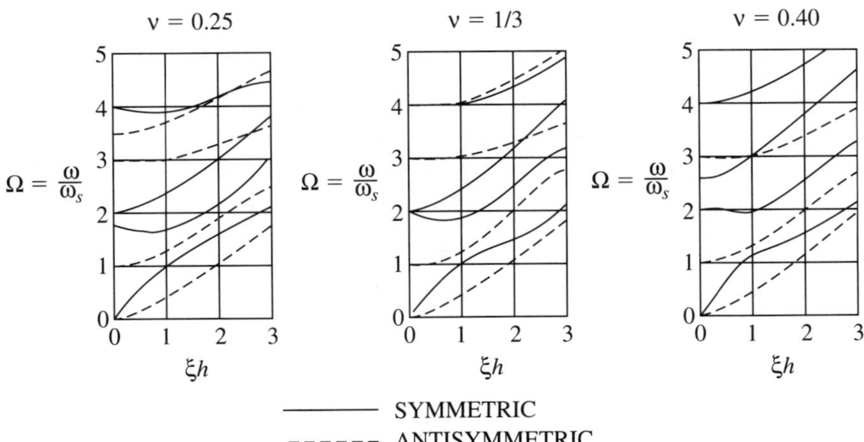

Fig. 2.2.2. Frequency spectrum of an infinite plate for small real wave number (after Mindlin).

2.2.2 Lowest Branches of Frequency Spectrum

When both the frequency ω and wave number ξ are small, the wave numbers α and β also will be small according to Eqs. (2.2.4), and we may introduce the approximations

$$\tan \alpha h \approx \alpha h \left(1 + \tfrac{1}{3}\alpha^2 h^2\right)$$
$$\tan \beta h \approx \beta h \left(1 + \tfrac{1}{3}\beta^2 h^2\right).$$

Substitution of these into the frequency equations (2.2.7) and (2.2.8) yields, respectively,

$$\omega = \xi \sqrt{\frac{E}{\rho(1 - \nu^2)}} \tag{2.2.9}$$

$$\omega = \xi^2 h \sqrt{\frac{E}{3\rho(1 - \nu^2)}}. \tag{2.2.10}$$

These are the results for an infinite plate that are also given by the classical plate theories for extension and flexure, respectively, as will be discussed in the next chapter.

2.2.3 Cutoff Frequencies at $\xi = 0$

Corresponding to an infinitely long wavelength, the wave number ξ is equal to 0, and the frequencies of the various branches of the spectrum become the cutoff frequencies. By means of a limiting process, Mindlin showed that the cutoff frequencies given by the Rayleigh–Lamb solution in Eqs. (2.2.7) and (2.2.8) are identical with the frequencies of simple thickness modes of vibration of an infinite plate, which will be discussed in detail in the next section.

2.2.4 Rayleigh Surface Waves at Large ξ

As the wavelength becomes very short, the wave number ξ becomes very large, and the various branches of the frequency spectrum approach asymptotically straight lines passing through the origin, that is, the phase velocity becomes constant. In fact, we have

$$\frac{\tan \beta h}{\tan \alpha h} \to 1 \quad \text{as} \quad \xi h \to \infty,$$

and Eqs. (2.2.7) and (2.2.8) both reduce to

$$(\xi^2 - \beta^2)^2 = -4\xi^2 \alpha \beta$$

or, by squaring both sides,

$$\left(\frac{\omega^2}{\xi^2 c_\beta^2} - 2\right)^4 = 16 \left(\frac{\omega^2}{\xi^2 c_\alpha^2} - 1\right)\left(\frac{\omega^2}{\xi^2 c_\beta^2} - 1\right), \tag{2.2.11}$$

which is the same as the governing equation of the phase velocity ($c = \omega/\xi$) of Rayleigh's surface waves. In the range of $c < c_\beta$, this equation has only one root, which is the situation with $\nu = 1/3$ for which $c = 0.932c_\beta$. The value of the phase velocity given by Eq. (2.2.11) is the value of the slope of the asymptotic line for the two lowest branches, corresponding to the lowest symmetric and antisymmetric modes. All higher branches approach the asymptotic line with a slope equal to $c = c_\beta$.

2.2.5 Finite Plates

As was mentioned earlier, the complexity of the theory of vibrations of plates is due to the fact that the two types of stress waves, dilatational and rotational, become coupled with each other if any part of the plate boundary is free of traction. Mindlin (1960) developed an interesting analysis of such coupling by tracing the two types of waves systematically from the infinite body to the half-space, then to the infinite plate, and, finally, to the finite plate. The process is made continuous by first considering each new plane boundary with mixed boundary conditions (combining one condition with prescribed displacement and another with prescribed traction) so that no coupling between the two types of waves takes place. A plane boundary with variable elastic restraint is introduced next to allow coupling, and this finally passes into a free boundary. Through such an analysis, Mindlin was able to explain the terrace-like structure of the frequency spectrum associated with the higher modes of a free plate that has been observed in experiments with anisotropic resonators. As a practical matter, he reached the important conclusion that the adequacy of approximate plate equations for the prediction of natural frequencies of finite plates, within a frequency range, depends on the ability of the plate equations to reproduce accurately the frequency spectrum of an infinite plate, for that frequency range, as compared with the result of the exact elasticity theory.

2.3   Simple Thickness Modes in an Infinite Plate

Simple thickness modes are modes of free vibration in an infinite plate in which the boundary planes of the plate are traction-free and the displacements are dependent upon only the thickness coordinate. There are two types of simple thickness modes: thickness-stretch and thickness-shear. The simple thickness-stretch mode involves displacements in the thickness direction, and the simple thickness-shear involves those parallel to the plane of the plate. For an isotropic plate, the two types of simple thickness modes are uncoupled from each other.

For the simple thickness modes in an isotropic infinite plate, the governing equations of motion are deduced from Eqs. (2.1.3):

$$(\lambda + 2\mu)\frac{\partial^2 u_z}{\partial z^2} = \rho\frac{\partial^2 u_z}{\partial t^2}, \quad \mu\frac{\partial^2 u_x}{\partial z^2} = \rho\frac{\partial^2 u_x}{\partial t^2}.$$

The general solutions of these equations are

$$u_z = (A \sin \alpha z + B \cos \alpha z)\, e^{i\omega t}$$
$$u_x = (C \sin \beta z + D \cos \beta z)\, e^{i\omega t}. \tag{2.3.1}$$

The wave numbers are now

$$\alpha = \frac{\omega}{c_\alpha}, \quad \beta = \frac{\omega}{c_\beta}, \tag{2.3.2}$$

which are special cases of Eqs. (2.2.4) for $\xi = 0$. The boundary conditions for free tractions are

$$\sigma_{zz} = \sigma_{zx} = 0 \quad \text{at} \quad z = \pm h,$$

from which

$$\frac{\partial u_z}{\partial z} = \frac{\partial u_x}{\partial z} = 0 \quad \text{at} \quad z = \pm h.$$

By virtue of Eqs. (2.3.1), these conditions yield

$$A\alpha \cos ah \mp B\alpha \sin \alpha h = 0 \tag{2.3.3}$$
$$C\beta \cos \beta h \mp D\beta \sin \beta h = 0. \tag{2.3.4}$$

For simple thickness-stretch modes, we find from Eq. (2.3.3)

$$B = 0, \quad \alpha = \frac{p\pi}{2h} \quad \text{for} \quad p = \text{odd}$$

or

$$A = 0, \quad \alpha = \frac{p\pi}{2h} \quad \text{for} \quad p = \text{even}.$$

The frequency is therefore, according to Eqs. (2.3.2),

$$\omega = c_\alpha \frac{p\pi}{2h}, \tag{2.3.5}$$

and the displacement is, with a factor $e^{i\omega t}$ omitted,

$$u_z = A \sin \frac{p\pi z}{2h} \quad \text{for} \quad p = \text{odd}$$
$$u_z = B \cos \frac{p\pi z}{2h} \quad \text{for} \quad p = \text{even}, \tag{2.3.6}$$

which are symmetric and antisymmetric, respectively, with respect to the middle plane of the plate.

For simple thickness-shear modes, we find similarly from Eqs. (2.3.4),

$$\omega = c_\beta \frac{q\pi}{2h} \tag{2.3.7}$$

and

$$u_x = C \sin \frac{q\pi z}{2h} \quad \text{for} \quad q = \text{odd}$$
$$u_x = D \cos \frac{q\pi z}{2h} \quad \text{for} \quad q = \text{even}, \tag{2.3.8}$$

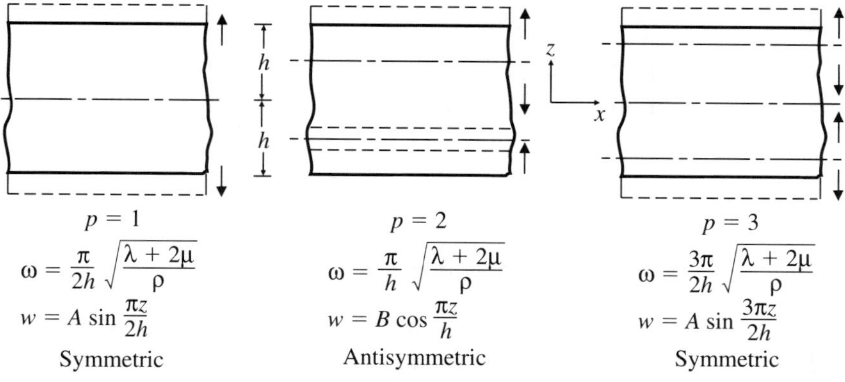

$p = 1$

$$\omega = \frac{\pi}{2h} \sqrt{\frac{\lambda + 2\mu}{\rho}}$$

$$w = A \sin \frac{\pi z}{2h}$$

Symmetric

$p = 2$

$$\omega = \frac{\pi}{h} \sqrt{\frac{\lambda + 2\mu}{\rho}}$$

$$w = B \cos \frac{\pi z}{h}$$

Antisymmetric

$p = 3$

$$\omega = \frac{3\pi}{2h} \sqrt{\frac{\lambda + 2\mu}{\rho}}$$

$$w = A \sin \frac{3\pi z}{2h}$$

Symmetric

Fig. 2.3.1. Simple thickness-stretch modes (after Mindlin)

which are symmetric and antisymmetric, respectively, with respect to the middle plane of the plate.

The frequencies and shapes of the first three simple thickness-stretch and first three simple thickness-shear modes, as given by Mindlin (1955), are shown in Figures 2.3.1 and 2.3.2, respectively. In these results, p or q identifies the number of nodal planes parallel to the boundary planes of the plate. The antisymmetric mode of the lowest frequency is always the first thickness-shear mode $q = 1$. The next higher antisymmetric frequency is either that of the second antisymmetric thickness-shear mode $q = 3$, or the first antisymmetric thickness-stretch mode $p = 2$, depending on whether the Poisson ratio is greater or smaller than $1/10$. In any case, the second antisymmetric frequency is between $2\sqrt{2}$ and 3 times the first. The symmetric mode of the lowest frequency may be either the first symmetric thickness-stretch mode $p = 1$, or the first symmetric thickness-shear mode $q = 2$, according to whether the Poisson ratio is smaller or greater than $1/3$. Coincidence of frequencies of simple thickness modes occurs when

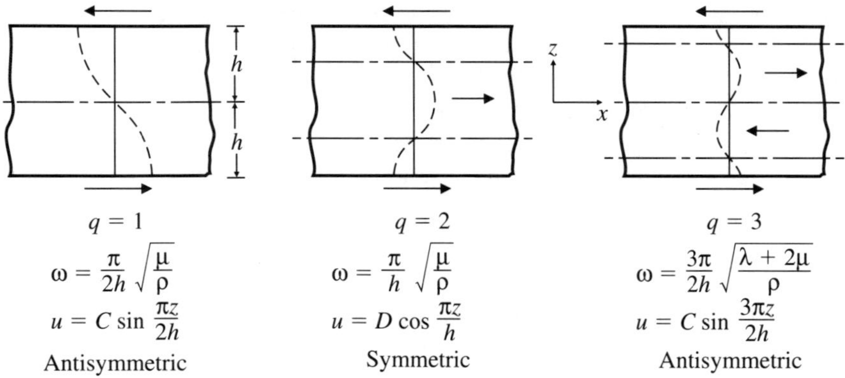

$q = 1$

$$\omega = \frac{\pi}{2h} \sqrt{\frac{\mu}{\rho}}$$

$$u = C \sin \frac{\pi z}{2h}$$

Antisymmetric

$q = 2$

$$\omega = \frac{\pi}{h} \sqrt{\frac{\mu}{\rho}}$$

$$u = D \cos \frac{\pi z}{h}$$

Symmetric

$q = 3$

$$\omega = \frac{3\pi}{2h} \sqrt{\frac{\lambda + 2\mu}{\rho}}$$

$$u = C \sin \frac{3\pi z}{2h}$$

Antisymmetric

Fig. 2.3.2 Simple thickness-shear modes (after Mindlin)

$$p^2(\lambda + 2\mu) = q^2\mu \quad \text{or} \quad \nu = \frac{q^2 - 2p^2}{2(q^2 - p^2)}. \qquad (2.3.9)$$

We next consider Mindlin's limiting process mentioned in Section 2.2. The process starts with the Rayleigh–Lamb solution in Eqs. (2.2.7) and (2.2.8) by taking the frequency in the form

$$\omega = (p + \epsilon)\frac{c_\alpha \pi}{2h}$$
$$\omega = (q + \epsilon)\frac{c_\beta \pi}{2h} \qquad (2.3.10)$$

with $|\epsilon| \ll 1$ in the neighborhood of $\xi \ll 1$. This is to obtain the results in Eqs. (2.3.5) and (2.3.7) in the limiting case of $\xi \to 0$ and $\epsilon \to 0$. The limiting slopes and curvatures of the branches may be evaluated similarly. Mindlin found that, in general, at $\xi = 0$, the slopes are 0 and the curvatures are positive or negative according to

$$\frac{p\pi}{4} \pm \frac{4}{c^3}\cot\frac{p\pi}{2} > \text{or} < 0$$
$$\frac{q\pi}{4} \mp \frac{4}{c}\tan\frac{q\pi}{2c} > \text{or} < 0,$$

where the upper and lower signs between the first two terms in each line apply to symmetric and antisymmetric modes, respectively, and

$$c = \frac{c_\alpha}{c_\beta} = \sqrt{\frac{\lambda + 2\mu}{\mu}} = \sqrt{\frac{2(1 - \nu)}{1 - 2\nu}}. \qquad (2.3.11)$$

The exception to these general rules is in the cases in which

$$c = \frac{q}{p} \quad \text{or} \quad \nu = \frac{q^2 - 2p^2}{2(q^2 - p^2)}. \qquad (2.3.12)$$

This is the same as Eq. (2.3.9) for coincidence. Whenever this relation is true, not only are the frequencies of the two modes denoted by p and q the same, but also the slopes of the two branches at $\xi = 0$ are no longer 0, and the corresponding curvatures become infinite. All of these conclusions have been confirmed by the results in Figure 2.2.2. In the particular case of $\nu = 1/3$, the cutoff frequency is the same for $p = 1$ and $q = 2$, and the values of ν, p, and q satisfy Eq. (2.3.9) or Eq. (2.3.12).

2.4   Horizontal Shear Modes in an Infinite Plate

The horizontal shear modes are associated with a single displacement component in one direction that is a function of the rectangular coordinates in the other two directions. They can be put in the typical form

$$u_x = u_x(y, z, t), \quad u_y = u_z = 0.$$

Of the four potentials in Eqs. (2.1.4), only the following is needed:

$$H_3 = (C \sin \beta z + D \cos \beta z) \sin \eta y \, e^{i\omega t}. \tag{2.4.1}$$

Since H_3 must satisfy the second of Eqs. (2.1.7), we have

$$\beta^2 + \eta^2 = \frac{\omega^2}{c_\beta^2}. \tag{2.4.2}$$

The nonzero displacement and stresses are then

$$u_x = \frac{\partial H_3}{\partial y} = \eta(C \sin \beta z + D \cos \beta z) \cos \eta y \, e^{i\omega t}$$

$$\sigma_{xz} = \mu \frac{\partial u_x}{\partial z} = \mu \beta \eta (C \cos \beta z - D \sin \beta z) \cos \eta y \, e^{i\omega t} \tag{2.4.3}$$

$$\sigma_{xy} = \mu \frac{\partial u_x}{\partial y} = -\mu \eta^2 (C \sin \beta z + D \cos \beta z) \sin \eta y \, e^{i\omega t}.$$

For an infinite plate with traction-free boundary planes, the boundary conditions are

$$\sigma_{xz} = 0 \quad \text{at} \quad z = \pm h.$$

These require that, according to the second of Eqs. (2.4.3),

$$C \cos \beta h \mp D \sin \beta h = 0,$$

from which we find

$$D = 0, \quad \beta = \frac{q\pi}{2h} \quad \text{for} \quad q = \text{odd}$$

or

$$C = 0, \quad \beta = \frac{q\pi}{2h} \quad \text{for} \quad q = \text{even}.$$

These correspond to modes antisymmetric or symmetric with respect to the middle plane of the plate. With this value of β substituted into Eq. (2.4.2), we find

$$\left(\frac{\omega}{\omega_s}\right)^2 = q^2 + \left(\frac{2\eta h}{\pi}\right)^2 \quad (q = 1, 2, 3 \ldots). \tag{2.4.4}$$

The frequencies therefore increase monotonically with η. Furthermore, the cutoff frequencies of the horizontal shear modes at $\eta = 0$ are the same as the simple thickness-shear frequencies.

2.5   Modes in an Infinite Plate Involving Phase Reversals in Both x- and y-Directions

The Rayleigh–Lamb solution in Section 2.2 may readily be extended to include phase reversals that occur similtaneously in both the x- and y-directions, by taking the potentials in the form

$$\phi = f(z) \sin \xi x \sin \eta y \, e^{i\omega t}$$
$$H_1 = -g_1(z) \sin \xi x \cos \eta y \, e^{i\omega t}$$
$$H_2 = -g_2(z) \cos \xi x \sin \eta y \, e^{i\omega t} \tag{2.5.1}$$
$$H_3 = -g_3(z) \cos \xi x \cos \eta y \, e^{i\omega t}.$$

By a procedure similar to that used before, the frequency equations are found to be, for symmetric and antisymmetric modes, respectively,

$$\frac{\tan \beta h}{\tan \alpha h} = -\frac{4\alpha\beta(\xi^2 + \eta^2)}{(\xi^2 + \eta^2 - \beta^2)^2} \tag{2.5.2}$$

$$\frac{\tan \beta h}{\tan \alpha h} = -\frac{(\xi^2 + \eta^2 - \beta^2)^2}{4\alpha\beta(\xi^2 + \eta^2)}. \tag{2.5.3}$$

These are a generalization of Eqs. (2.2.7) and (2.2.8), respectively, with ξ^2 replaced by $(\xi^2 + \eta^2)$, as pointed out by Rayleigh.

In a similar manner, the horizontal shear modes in Section 2.4 may be generalized to include phase reversals in both the x- and y-directions. The potential H_3 in Eq. (2.4.1) is now replaced by

$$H_3 = (C \sin \beta z + D \cos \beta z) \sin \xi x \sin \eta y \, e^{i\omega t}$$

with

$$\beta^2 + \xi^2 + \eta^2 = \frac{\omega^2}{c_\beta^2}.$$

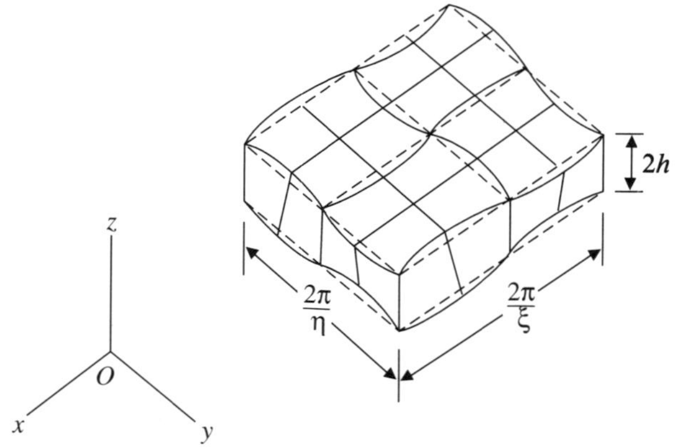

Fig. 2.5.1. Thickness-twist mode.

The frequency is then given by

$$\left(\frac{\omega}{\omega_s}\right)^2 = q^2 + \left(\frac{2h}{\pi}\right)^2 (\xi^2 + \eta^2) \quad (q = 1, 2, 3, \ldots), \tag{2.5.4}$$

which is a generalization of Eq. (2.4.4). The corresponding mode is a thickness-twist mode, as shown in Figure 2.5.1.

2.6 Plane-Strain Modes in an Infinite Sandwich Plate

We consider an isotropic elastic sandwich plate that has three layers symmetrically constructed about the middle plane of the sandwich, with two identical face layers. Perfect bonds are assumed to exist at the interfaces between the face and core layers. Because of the symmetrical construction, symmetric and antisymmetric modes of vibration of the sandwich plate are again uncoupled from each other just as in the case of a single-layered plate. The symmetric sandwich therefore represents the simplest layered type. On the other hand, no restrictions are imposed upon the magnitudes of the ratios between the thicknesses, material densities, and elastic constants of the core and face layers. This three-layered sandwich, which was first formulated and investigated by Yu (1959a,b, 1960), is therefore more general than, and includes as a special case, an ordinary sandwich plate that has a thick but light and soft core layer.

As is shown in Figure 2.6.1, the middle plane of the sandwich plate is chosen to be the xy-plane. In the z-direction, the thicknesses of the lower face, core, and upper face layers extend from $-h$ to $-h_1$, $-h_1$ to h_1, and h_1 to h, respectively. The thickness of the core is thus $2h_1$, and that of each of the two face layers is $h_2 = h - h_1$. To identify the layers, a subscript i will be used; the values of $i = 1, 2$, and 3 denote the core, lower face, and upper face layers, respectively. When only the values 1 and 2 are used for the subscript, they refer to the core layer and face layers, respectively.

For plane-strain modes in an infinite sandwich plate, the general solution in Eqs. (2.2.6) for a single-layered plate is still applicable to each of the three individual layers. By inserting a subscript i, with $i = 1, 2$, and 3, the displacements and stresses needed in formulating the boundary conditions are written as

$$\begin{aligned}
u_{xi} &= (\xi f_i + g_i') \cos \xi x \, e^{i\omega t} \\
u_{zi} &= (f_i' + \xi g_i) \sin \xi x \, e^{i\omega t} \\
\sigma_{zzi} &= -\mu_i[(\beta_i^2 - \xi^2) f_i - 2\xi g_i'] \sin \xi x \, e^{i\omega t} \\
\sigma_{zxi} &= \mu_i[2\xi f_i' + (\xi^2 - \beta_i^2) g_i] \cos \xi x \, e^{i\omega t},
\end{aligned} \tag{2.6.1}$$

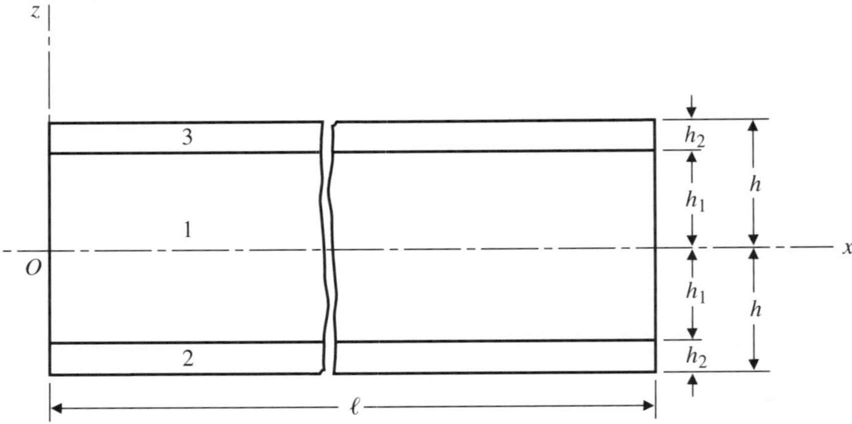

Fig. 2.6.1. Cross-section of sandwich plate.

where

$$\alpha_i^2 = \frac{\omega^2}{c_{\alpha i}^2} - \xi^2, \qquad \beta_i^2 = \frac{\omega^2}{c_{\beta i}^2} - \xi^2$$

$$c_{\alpha i}^2 = \frac{\lambda_i + 2\mu_i}{\rho_i}, \qquad c_{\beta i}^2 = \frac{\mu_i}{\rho_i}$$

(2.6.2)

and

$$f_i = A_i \sin \alpha_i z + B_i \cos \alpha_i z$$

$$g_i = C_i \sin \beta_i z + D_i \cos \beta_i z.$$

(2.6.3)

The conditions of zero traction still are applied to the boundary planes $z = \pm h$ of the sandwich plate. In addition, under the assumption of a perfect bond at the interface planes $z = \pm h_1$, the displacements and stresses at the interfaces in adjacent layers are required to be continuous.

For antisymmetric modes, f_i and g_i are taken in the following forms:

$$f_1 = A_1 \sin \alpha_1 z$$

$$\left.\begin{matrix} f_2 \\ f_3 \end{matrix}\right\} = A_2 \sin \alpha_2 z \pm B_2 \cos a_2 z$$

$$g_1 = D_1 \cos \beta_1 z$$

$$\left.\begin{matrix} g_2 \\ g_3 \end{matrix}\right\} = \pm C_2 \sin \beta_2 z + D_2 \cos \beta_2 z.$$

(2.6.4)

The boundary conditions are therefore

$$\sigma_{zx2} = \sigma_{zz2} = 0 \quad \text{at } z = -h$$

$$\sigma_{zx3} = \sigma_{zz3} = 0 \quad \text{at } z = h$$

$$\left.\begin{matrix} u_{x2} = u_{x1}, \quad u_{z2} = u_{z1} \\ \sigma_{zx2} = \sigma_{zx1}, \ \sigma_{zz2} = \sigma_{zz1} \end{matrix}\right\} \text{at } z = -h_1$$

$$\left.\begin{matrix} u_{x3} = u_{x1}, \quad u_{z3} = u_{z1} \\ \sigma_{zx3} = \sigma_{zx1}, \ \sigma_{zz3} = \sigma_{zz1} \end{matrix}\right\} \text{at } z = h_1.$$

(2.6.5)

With f_i and g_i substituted from Eqs. (2.6.4) into (2.6.1) and (2.6.5), we find six homogeneous equations linear in the six constants A_1, D_1, A_2, B_2, C_2, and D_2. When the determinant of the coefficients of the six constants is set equal to zero in the usual manner, the following frequency equation is obtained (Yu 1960):

$$
\begin{aligned}
8\xi^2 \, &\alpha_1 \, \alpha_2\beta_1\beta_2(r_\mu - 1)(\xi^2 - \beta_2^2)[2\xi^2 - r_\mu(\xi^2 - \beta_2^2)]\cos \alpha_1 h_1 \sin \beta_1 h_1 \\
&+ 8\xi^2\alpha_2\beta_2(\xi^2 - \beta_2^2)[2r_\mu\xi^2 - (\xi^2 - \beta_1^2)] \\
&\quad\times[(\xi^2 - \beta_1^2) - r_\mu(\xi^2 - \beta_2^2)] \sin \alpha_1 h_1 \cos \beta_1 h_1 \\
&+ \{[2\xi^2 - r_\mu(\xi^2 - \beta_2^2)]^2\alpha_1\beta_1 \cos \alpha_1 h_1 \sin \beta_1 h_1 \\
&\quad +[(\xi^2 - \beta_1^2) - r_\mu(\xi^2 - \beta_2^2)]^2\xi^2 \sin \alpha_1 h_1 \cos \beta_1 h_1\} \\
&\quad\times[4\xi^2\alpha_2\beta_2 \cos \alpha_2 h_2 \cos \beta_2 h_2 - (\xi^2 - \beta_2^2)^2 \sin \alpha_2 h_2 \sin \beta_2 h_2] \\
&- \{4(r_\mu - 1)^2\xi^2\alpha_1\alpha_2\beta_1\beta_2 \cos \alpha_1 h_1 \sin \beta_1 h_1 \\
&\quad +[2r_\mu\xi^2 - (\xi^2 - \beta_1^2)]^2\alpha_2\beta_2 \sin \alpha_1 h_1 \cos \beta_1 h_1\} \\
&\quad\times[4\xi^2\alpha_2\beta_2 \sin \alpha_2 h_2 \sin \beta_2 h_2 - (\xi^2 - \beta_2^2)^2 \cos \alpha_2 h_2 \cos \beta_2 h_2] \\
&+ r_\mu\alpha_1\beta_2(\xi^2 + \beta_1^2)(\xi^2 + \beta_2^2) \cos \alpha_1 h_1 \cos \beta_1 h_1 \\
&\quad\times[4\xi^2\alpha_2\beta_2 \cos \alpha_2 h_2 \sin \beta_2 h_2 - (\xi^2 - \beta_2^2)^2 \sin \alpha_2 h_2 \cos \beta_2 h_2] \\
&- r_\mu\alpha_2\beta_1(\xi^2 + \beta_1^2)(\xi^2 + \beta_2^2) \sin \alpha_1 h_1 \sin \beta_1 h_1 \\
&\quad\times[4\xi^2\alpha_2\beta_2 \sin \alpha_2 h_2 \cos \beta_2 h_2 - (\xi^2 - \beta_2^2)^2 \cos \alpha_2 h_2 \sin \beta_2 h_2] \\
&= 0,
\end{aligned}
\tag{2.6.6}
$$

where $r_\mu = \mu_2/\mu_1$.

Similarly, for symmetric modes, f_i and g_i are taken in the forms

$$f_1 = B_1 \cos \alpha_1 z$$

$$\left.\begin{aligned} f_2 \\ f_3 \end{aligned}\right\} = \pm A_2 \sin \alpha_2 z + B_2 \cos \alpha_2 z$$

$$g_1 = C_1 \sin \beta_1 z$$

$$\left.\begin{aligned} g_2 \\ g_3 \end{aligned}\right\} = C_2 \sin \beta_2 z \pm D_2 \cos \beta_2 z.$$

The frequency equation is similar to Eq. (2.6.6) and obtainable from the latter by making appropriate replacements of the trigonometric functions as follows (Yu 1962):

Eq. (2.6.6) with $\sin \alpha_1 h_1$, $\cos \alpha_1 h_1$, $\sin \beta_1 h_1$, $\cos \beta_1 h_1$ replaced by

$$\cos \alpha_1 h_1, \ -\sin \alpha_1 h_1, \ -\cos \beta_1 h_1, \ \sin \beta_1 h_1, \text{ respectively.} \tag{2.6.7}$$

By taking $h_1 = 0$ or $h_2 = r_\mu = 0$, Eqs. (2.6.6) and (2.6.7) reduce to the Rayleigh–Lamb solutions for a single-layered homogeneous plate.

In spite of the complexity, the frequency equations (2.6.6) and (2.6.7) derived from the elasticity theory have been applied to numerical examples. These are to be discussed in Chapter 4, where comparison will be made with corresponding results given by sandwich plate equations to be developed. In this chapter, we shall cover next a detailed analysis of the simple thickness modes in an infinite sandwich plate. The importance of a thorough understanding of such modes has just been demonstrated for a single-layered plate. Additional complexity for a sandwich plate will still be manageable.

2.7   Simple Thickness Modes in an Infinite Sandwich Plate

For simple thickness modes in an infinite sandwich plate (Yu 1995), the wave number ξ is 0, for which the frequency equations (2.6.6) and (2.6.7) for plane-strain modes reduce to those for simple thickness modes in addition to providing 0 frequency values for the lowest branches of the frequency spectrum. In this section, we calculate directly the frequencies and mode shapes of the simple thickness modes of a sandwich by generalizing the treatment for a single-layered plate in Section 2.3. Thus, by inserting a subscript i, with $i = 1, 2$, and 3, the general solution in Eqs. (2.3.1) and (2.3.2) takes the form

$$u_{zi} = (A_i \sin \alpha_i z + B_i \cos \alpha_i z) e^{i\omega t}$$
$$u_{xi} = (C_i \sin \beta_i z + D_i \cos \beta_i z) e^{i\omega t},$$

where

$$\alpha_i = \frac{\omega}{c_{\alpha i}}, \quad \beta_i = \frac{\omega}{c_{\beta i}}.$$

The boundary conditions for the simple thickness modes are the same as those for plane-strain modes: vanishing tractions at the boundary planes $z = \pm h$ and continuous displacements u_{zi}, u_{xi} and stresses σ_{zxi}, σ_{zzi} at the interface planes $z = \pm h_1$.

The simple thickness-stretch and thickness-shear modes are again uncoupled, as in the case of a single-layered plate. By substituting the general solution into the boundary conditions, the frequencies and modes are determined in the usual manner. The results for the various types of simple thickness modes are as follows:

(a) Symmetric thickness-stretch modes:

$$\tan \alpha_1 h_1 \tan \alpha_2 h_2 = \frac{(\lambda_1 + 2\mu_1)\alpha_1}{(\lambda_2 + 2\mu_2)\alpha_2}$$

$$= \sqrt{\frac{(\lambda_1 + 2\mu_1)\rho_1}{(\lambda_2 + 2\mu_2)\rho_2}}$$

$$u_{z1} = A_1 \sin \alpha_1 z \qquad\qquad (2.7.1a,b,c)$$

$$\left.\begin{array}{c} u_{z2} \\ u_{z3} \end{array}\right\} = \mp A_1 \frac{\sin \alpha_1 h_1}{\cos \alpha_2 h_2} \cos \alpha_2 (h \pm z).$$

(b) Antisymmetric thickness-stretch modes:

$$\frac{\tan \alpha_2 h_2}{\tan \alpha_1 h_1} = -\frac{(\lambda_1 + 2\mu_1)\alpha_1}{(\lambda_2 + 2\mu_2)\alpha_2}$$

$$= -\sqrt{\frac{(\lambda_1 + 2\mu_1)\rho_1}{(\lambda_2 + 2\mu_2)\rho_2}} \qquad (2.7.2a,b,c)$$

$$u_{z1} = B_1 \cos \alpha_1 z$$

$$\left.\begin{array}{c} u_{z2} \\ u_{z3} \end{array}\right\} = B_1 \frac{\cos \alpha_1 h_1}{\cos \alpha_2 h_2} \cos \alpha_2 (h \pm z).$$

(c) Antisymmetric thickness-shear modes:

$$\tan \beta_1 h_1 \tan \beta_2 h_2 = \frac{\mu_1 \beta_1}{\mu_2 \beta_2} = \sqrt{\frac{\mu_1 \rho_1}{\mu_2 \rho_2}}$$

$$u_{x1} = C_1 \sin \beta_1 z \qquad (2.7.3a,b,c)$$

$$\left.\begin{array}{c} u_{x2} \\ u_{x3} \end{array}\right\} = \mp C_1 \frac{\sin \beta_1 h_1}{\cos \beta_2 h_2} \cos \beta_2 (h \pm z).$$

(d) Symmetric thickness-shear modes:

$$\frac{\tan \beta_2 h_2}{\tan \beta_1 h_1} = -\frac{\mu_1 \beta_1}{\mu_2 \beta_2} = -\sqrt{\frac{\mu_1 \rho_1}{\mu_2 \rho_2}}$$

$$u_{x1} = D_1 \cos \beta_1 z \qquad (2.7.4a,b,c)$$

$$\left.\begin{array}{c} u_{x2} \\ u_{x3} \end{array}\right\} = D_1 \frac{\cos \beta_1 h_1}{\cos \beta_2 h_2} \cos \beta_2 (h \pm z).$$

A factor $e^{i\omega t}$ has been omitted in all of the above expressions of displacements.

It will be convenient to introduce the dimensionless frequencies f and f' such that we may write

$$\beta_1 h_1 = \omega h_1 \sqrt{\frac{\rho_1}{\mu_1}} = \frac{\pi \omega}{2\omega_{s1}} = f$$

$$\alpha_1 h_1 = \omega h_1 \sqrt{\frac{\rho_1}{\lambda_1 + 2\mu_1}} = \frac{\pi \omega}{2\omega_{s1}c_1} = \frac{f}{c_1} = f'$$

$$\beta_2 h_2 = f \sqrt{\frac{r_\rho r_h^2}{r_\mu}} \qquad (2.7.5)$$

$$\alpha_2 h_2 = f' \sqrt{\frac{r_\rho r_h^2}{r_{\lambda+2\mu}}}$$

where

$$\omega_{s1} = \sqrt{\frac{\pi}{2h_1} \frac{\mu_1}{\rho_1}} \qquad (2.7.6)$$

is the frequency of the lowest simple thickness-shear mode in a separate core layer, the parameter

$$c_1 = \sqrt{\frac{\lambda_1 + 2\mu_1}{\mu_1}} = \sqrt{\frac{2(1 - \nu_1)}{1 - 2\nu_1}} \qquad (2.7.7)$$

is always greater than 1, and the various ratios are

$$r_\rho = \frac{\rho_2}{\rho_1}, \quad r_h = \frac{h_2}{h_1}, \quad r_\mu = \frac{\mu_2}{\mu_1}, \quad r_{\lambda+2\mu} = \frac{\lambda_2 + 2\mu_2}{\lambda_1 + 2\mu_1}.$$

In terms of these dimensionless parameters and ratios, Eqs. (2.7.1) through (2.7.4) become

(a) Symmetric thickness-stretch modes:

$$\tan f' \tan \left(f' \sqrt{\frac{r_\rho r_h^2}{r_{\lambda+2\mu}}} \right) = \frac{1}{\sqrt{r_\rho r_{\lambda+2\mu}}}$$

$$u_{z1} = A_1 \sin \left(\frac{f'z}{h_1} \right) \qquad (2.7.8a,b,c)$$

$$\left. \begin{array}{c} u_{z2} \\ u_{z3} \end{array} \right\} = \mp A_1 \frac{\sin f'}{\cos \left(f' \sqrt{r_\rho r_h^2 / r_{\lambda+2\mu}} \right)} \cos \left(f' \sqrt{\frac{r_\rho r_h^2}{r_{\lambda+2\mu}}} \frac{h \pm z}{h_2} \right).$$

(b) Antisymmetric thickness-stretch modes:

$$\tan \left(f' \sqrt{\frac{r_\rho r_h^2}{r_{\lambda+2\mu}}} \right) \frac{1}{\tan f'} = -\frac{1}{\sqrt{r_\rho r_{\lambda+2\mu}}}$$

$$u_{z1} = B_1 \cos \left(\frac{f'z}{h_1} \right) \qquad (2.7.9a,b,c)$$

$$\left. \begin{array}{c} u_{z2} \\ u_{z3} \end{array} \right\} = B_1 \frac{\cos f'}{\cos \left(f' \sqrt{r_\rho r_h^2 / r_{\lambda+2\mu}} \right)} \cos \left(f' \sqrt{\frac{r_\rho r_h^2}{r_{\lambda+2\mu}}} \frac{h \pm z}{h_2} \right).$$

(c) Antisymmetric thickness-shear modes:

$$\tan f \tan \left(f \sqrt{\frac{r_\rho r_h^2}{r_\mu}} \right) = \frac{1}{\sqrt{r_\rho r_\mu}}$$

$$u_{x1} = C_1 \sin \left(\frac{fz}{h_1} \right) \qquad (2.7.10a,b,c)$$

$$\left. \begin{array}{c} u_{x2} \\ u_{x3} \end{array} \right\} = \mp C_1 \frac{\sin f}{\cos \left(f \sqrt{r_\rho r_h^2 / r_\mu} \right)} \cos \left(f \sqrt{\frac{r_\rho r_h^2}{r_\mu}} \frac{h \pm z}{h_2} \right).$$

(d) Symmetric thickness-shear modes:

$$\tan\left(f\sqrt{\frac{r_\rho r_h^2}{r_\mu}}\right)\frac{1}{\tan f} = -\frac{1}{\sqrt{r_\rho r_\mu}}$$

$$u_{x1} = D_1 \cos\left(\frac{fz}{h_1}\right)$$

$$\left.\begin{array}{c} u_{x2} \\ u_{x3} \end{array}\right\} = D_1 \frac{\cos f}{\cos\left(f\sqrt{r_\rho r_h^2/r_\mu}\right)} \cos\left(f\sqrt{\frac{r_\rho r_h^2}{r_\mu}}\frac{h\pm z}{h_2}\right).$$

$$(2.7.11a,b,c)$$

The mode shapes are thus still harmonic in the thickness direction as in the case of a single-layered plate. However, while there are always an integer number of half-waves across the thickness in the single-layered case, it is not necessarily so in the sandwich case.

Up to this point, the results have been valid for an arbitrary symmetric three-layered sandwich. We shall next restrict our attention to ordinary sandwich plates that have a relatively thick but light and soft core layer for which r_h is usually small, r_ρ large, and r_μ and $r_{\lambda+2\mu}$ very large, as compared with unity, so that

$$\frac{r_\rho r_h^2}{r_{\lambda+2\mu}}, \quad \frac{r_\rho r_h^2}{r_\mu} \ll 1.$$

For the lowest simple thickness modes, the dimensionless frequencies f and f' are further assumed to be small enough so that the tangents of the angles $f'\sqrt{r_\rho r_h^2/r_{\lambda+2\mu}}$ and $f\sqrt{r_\rho r_h^2/r_\mu}$ may be replaced by the angles themselves and the cosines of these angles replaced by unity. Thus, Eqs. (2.7.8) through (2.7.11) reduce to

(a) Symmetric thickness-stretch modes:

$$f' \tan f' = \frac{1}{r_\rho r_h}$$

$$u_{z1} = A_1 \sin\frac{f'z}{h_1}$$

$$\left.\begin{array}{c} u_{z2} \\ u_{z3} \end{array}\right\} = \mp A_1 \sin f'.$$

$$(2.7.12a,b,c)$$

(b) Antisymmetric thickness-stretch modes:

$$\frac{f'}{\tan f'} = \frac{-1}{r_\rho r_h}$$

$$u_{z1} = B_1 \cos\frac{f'z}{h_1}$$

$$\left.\begin{array}{c} u_{z2} \\ u_{z3} \end{array}\right\} = \mp B_1 \cos f'.$$

$$(2.7.13a,b,c)$$

(c) Antisymmetric thickness-shear modes:

$$f \tan f = \frac{1}{r_\rho r_h}$$

$$u_{x1} = C_1 \sin \frac{fz}{h_1}$$

$$\left.\begin{array}{c} u_{x2} \\ u_{x3} \end{array}\right\} = \mp C_1 \sin f.$$

(2.7.14a,b,c)

(d) Symmetric thickness-shear modes:

$$\frac{f}{\tan f} = \frac{-1}{r_\rho r_h}$$

$$u_{x1} = D_1 \cos \frac{f'z}{h_1}$$

$$\left.\begin{array}{c} u_{x2} \\ u_{x3} \end{array}\right\} = \mp D_1 \cos f.$$

(2.7.15a,b,c)

We note first that the frequency equations (2.7.12a) through (2.7.15a) now involve only $r_\rho r_h$ and no longer $r_{\lambda+2\mu}$ or r_μ. Let f_a, f_b, f_c, and f_d be the lowest values of f given by these equations. Since the right sides of these equations are alternately positive and negative, we find

$$f_a' = \frac{f_a}{c_1} < \frac{\pi}{2}, \quad f_b' = \frac{f_b}{c_1} > \frac{\pi}{2}, \quad f_c < \frac{\pi}{2}, \quad f_d > \frac{\pi}{2},$$

from which

$$f_a < f_b, \quad f_c < f_d. \tag{2.7.16}$$

The lowest thickness-stretch mode is therefore symmetric, and the lowest thickness-shear mode is antisymmetric. Next, since $c_1 > 1$, we also have

$$f_c < f_b. \tag{2.7.17}$$

The antisymmetric mode with lowest frequency is therefore always the first thickness-shear mode, but the symmetric mode with the lowest frequency may be either the first symmetric thickness-stretch or the first symmetric thickness-shear mode. Finally, since the frequency equations for cases (a) and (c) have the same form, as do those for cases (b) and (d), we have

$$\frac{f_a}{c_1} = f_c, \quad \frac{f_b}{c_1} = f_d$$

and hence

$$f_a > f_c, \quad f_b > f_d. \tag{2.7.18}$$

According to the inequalities (2.7.16) through (2.7.18), the four frequencies now may be arranged in the following ascending order:

$$f_c < (f_a, f_d) < f_b. \tag{2.7.19}$$

The situation thus turns out to be entirely analogous to that with a single-layered plate.

In the case of a single-layered plate, whether f_a is greater or smaller than f_d depends on whether ν_1 is greater or smaller than 1/3. For a sandwich plate, this further depends on the value of $r_\rho r_h$. The results of f_a, f_b, f_c, and f_d calculated from the simplified frequency equations (2.7.12a) through (2.7.15a) are shown in Figure 2.7.1 for $\nu_1 = 1/4$, 1/3, and 1/2.5 and for $0 < r_\rho r_h < 2$. For $r_\rho r_h = 0$, the results are those of a single-layered plate. For very large $r_\rho r_h$, the results are no longer accurate because $r_\rho r_h^2/r_{\lambda_2\mu}$ and $r_\rho r_h/r_\mu$ were assumed to be small in the derivation of these equations. Results for $r_\rho r_h$ between 2 and 5 were available but have not been included. Nevertheless, results in Figure 2.7.1 are shown here for the first time.

For the case of $\nu_1 = 1/4$, the relation $f_a < f_d$ holds for the full range of $r_\rho r_h$. For the cases of $\nu_1 = 1/3$ and 1/2.5, the same relation is true only when $r_\rho r_h$ is greater than a limiting value to be determined by the point of intersection between

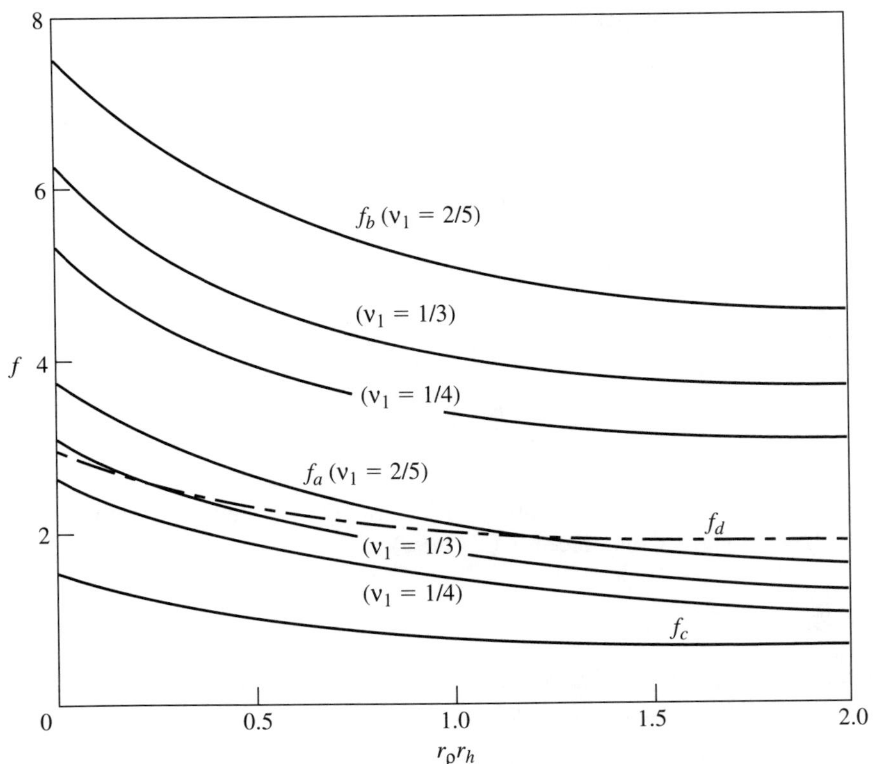

Fig. 2.7.1. Simple thickness frequencies of sandwich plates.

the f_a and f_d plots shown in Figure 2.7.1. The relative magnitudes of f_a and f_d are reversed when $r_\rho r_h$ is smaller than this limiting value. To determine the point of intersection, we first eliminate $r_\rho r_h$ from the frequency equations for cases (a) and (d), that is, Eqs. (2.7.12a) and (2.7.15a). This results in

$$\tan f \tan \frac{f}{c_1} = -c_1,$$

from which the lowest f may be determined for a given c_1 or ν_1. This is the common value of f_a and f_d at the point of intersection. The common value of $r_\rho r_h$ then may be determined from either Eq. (2.7.12a) or (2.7.15a).

The overall relations in the inequality (2.7.19) thus have been verified by the results in Figure 2.7.1. The results also show that f_a, f_b, f_c, and f_d all decrease monotonically with increasing $r_\rho r_h$, as they should according to the frequency equations (2.7.12a) through (2.7.15a).

According to Eqs. (2.7.12) through (2.7.15), the displacements for simple thickness modes in ordinary sandwich plates are approximately uniform across the thickness of each of the two face layers. For sufficiently low frequencies, the displacements in Eqs. (2.7.12b,c) and (2.7.14b,c) are further simplified into the following expressions:

(a) Symmetric thickness-stretch modes:

$$u_{z1} = A_1 \frac{f'z}{h_1}$$

$$\left.\begin{array}{c} u_{z2} \\ u_{z3} \end{array}\right\} = \mp A_1 f'.$$

(2.7.20)

(b) Antisymmetric thickness-shear modes:

$$u_{x1} = C_1 \frac{fz}{h_1}$$

$$\left.\begin{array}{c} u_{x2} \\ u_{x3} \end{array}\right\} = \mp C_1 f.$$

(2.7.21)

Results in Eqs. (2.7.21) were reported earlier (Yu 1959a), although those in Eqs. (2.7.20) were reported only recently (Yu 1995).

References

Deresiewicz, H., M.P. Bieniek, and F.L. DiMaggio, editors (1989) *The Collected Papers of Raymond D. Mindlin, Volumes I and II*. Springer-Verlag, New York.

Lamb, H. (1989) On Waves in an Elastic Plate. In: *Proceedings of London Mathematical Society*, Vol. 21, p. 85.

Mindlin, R.D. (1955) *An Introduction to the Mathematical Theory of Vibrations of Elastic Plates*. U.S. Army Signal Corps Engineering Laboratories, Fort Monmouth, New Jersey.

Mindlin, R.D. (1960) Waves and Vibrations in Isotropic, Elastic Plates. In: *Structural Mechanics*, pp. 199–232. Pergamon Press, New York.

Rayleigh, Lord. (1888) On the Free Vibrations of an Infinite Plate of Homogeneous Isotropic Elastic Matter. In: *Proceedings of London Mathematical Society*, Vol. 20, p. 225.

Yu, Y.Y. (1959a) A New Theory of Elastic Sandwich Plates—One-Dimensional Case. *Journal of Applied Mechanics*, Vol. 26, pp. 415–421.

Yu, Y.Y. (1959b) Simple Thickness-Shear Modes of Vibration of Infinite Sandwich Plates. *Journal of Applied Mechanics*, Vol. 26, pp. 679–681.

Yu, Y.Y. (1960) Flexural Vibrations of Elastic Sandwich Plates. *Journal of Aero/Space Sciences*, Vol. 27, pp. 272–283.

Yu, Y.Y. (1962) Extensional Vibrations of Elastic Sandwich Plates. In: *Proceedings of the Fourth U.S. National Congress of Applied Mechanics*, pp. 441–447.

Yu, Y.Y. (1995) Simple Thickness Modes of Vibration of a Sandwich Plate. Presented at Society of Engineering Science 32nd Annual Technical Meeting, New Orleans.

3

Linear Modeling of Homogeneous Plates

Vibration analysis of linear plates has been carried out in Chapter 2 on the basis of the exact elasticity theory. In this chapter, we again start from the elasticity theory, but proceed now to first derive linear equations of plates, which are then applied to the vibration analysis. As already mentioned, the plate equations are always approximate in nature from the standpoint of elasticity theory, but they are also always simpler to apply than the original elasticity equations.

Both classical and refined equations are treated here. In general, refined equations are more accurate and applicable to higher frequency ranges than the classical equations. In the case of flexure, the transverse shear effect was first included in the plate equations in the static case by Reissner (1945), and refined dynamic theories of beams and plates were introduced by Timoshenko (1921) and Mindlin (1951a), respectively, by including the effects of both transverse shear and rotatory inertia. Mindlin not only generalized the Timoshenko beam theory to the plate, but also compared and coordinated the results of his plate theory with those derived by Rayleigh and Lamb from elasticity theory described in the preceding chapter. This was one of the first attempts to establish a connection between the exact dynamic elasticity theory and an approximate plate theory, by means of which the shear factor in Mindlin's theory was determined. A refined linear theory for extension of an isotropic plate was first developed by Kane and Mindlin (1956).

In Sections 3.1 through 3.4 of this chapter, linear classical and refined equations for flexure (bending) and extension (stretching) of homogeneous plates will be derived from elasticity theory through systematic use of the linear variational equation of motion. While the procedure of deriving refined plate equations from the variational equation of motion is straightforward and now well-known, derivation of classical plate equations for flexure turns out to be rather involved and is apparently new (Yu 1992, 1995). With the classical and refined plate equations available, their ranges of usefulness in vibration analysis are examined in Section 3.5 by investigating the vibration of an infinite isotropic plate. To demonstrate the use of the generalized variational equation of motion in linear elasticity, we rederive (Yu 1965) a general system of linear equations of an anisotropic plate, originally given by Mindlin (1961), in the last section of this chapter.

3.1   Classical Equations for Flexure of a Homogeneous Plate

We consider a homogeneous, elastic, and isotropic plate with a uniform thickness $2h$, as shown in Figure 3.1.1. The middle plane of the undeformed plate is chosen to be the xy-plane, with boundary planes at $z = \pm h$. For simplicity, the plan area A of the plate is taken to be simply connected. The plate has a right cylindrical boundary surface S, which intersects the middle plane along a closed contour C. Directions normal and tangential to C are denoted by n and s, respectively.

The plate is considered to be thin in the usual sense that the thickness is small compared with dimensions in the plane of the plate. A typical derivation of the classical linear equations for flexure of a thin plate is given in the standard text by Timoshenko and Woinowsky-Krieger (1959). In this section, we make use of the variational equation of motion in three-dimensional linear elasticity theory (Yu 1992, 1995). By assuming a thickness dependence of the displacements and carrying out integration with respect to the thickness coordinate, the three-dimensional variational equation of motion becomes a two-dimensional one for the plate, from which the stress equations of motion and traction boundary conditions are readily derived.

3.1.1   Variational Equation of Motion

We start by assuming the three-dimensional displacements in the form

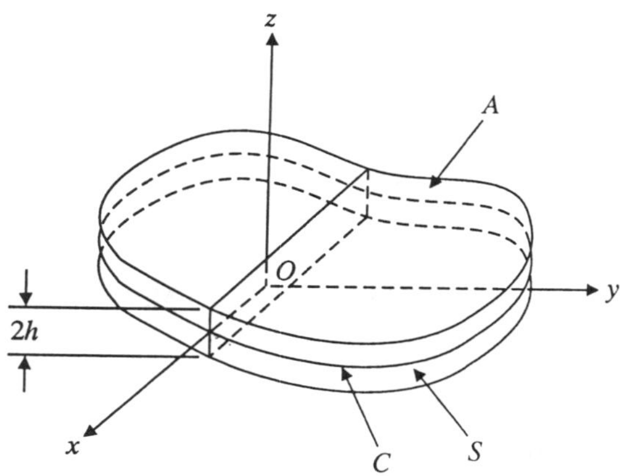

Fig. 3.1.1. A plate.

$$u_x(x, y, z, t) = -z\frac{\partial w}{\partial x}, \qquad u_y(x, y, z, t) = -z\frac{\partial w}{\partial y}, \tag{3.1.1}$$

$$u_z(x, y, z, t) = w(x, y, t),$$

where the deflection w is a plate displacement. Substituting these into the linear variational equation (1.4.9) and carrying out integration with respect to z over the plate thickness, we find

$$\int_{t_0}^{t_1} dt \int\int_A \left\{ \left[\frac{\partial M_x}{\partial x} + \frac{\partial M_{yx}}{\partial y} - Q_x + h(\sigma_{zx}^+ + \sigma_{zx}^-) \right. \right.$$

$$\left. + f_x^{(1)} + \rho\frac{2h^3}{3}\frac{\partial \ddot{w}}{\partial x} \right]\left[-\frac{\partial(\delta w)}{\partial x} \right]$$

$$+ \left[\frac{\partial M_{xy}}{\partial x} + \frac{\partial M_y}{\partial y} - Q_y + h(\sigma_{zy}^+ + \sigma_{zy}^-) \right.$$

$$\left. + f_y^{(1)} + \rho\frac{2h^3}{3}\frac{\partial \ddot{w}}{\partial y} \right]\left[-\frac{\partial(\delta w)}{\partial x} \right]$$

$$+ \left[\frac{\partial Q_x}{\partial x} + \frac{\partial Q_y}{\partial y} + (\sigma_{zz}^+ + \sigma_{zz}^-) \right.$$

$$\left. \left. + f_z^{(0)} - 2\rho h\ddot{w} \right]\delta w \right\} dx\,dy$$

$$- \int_{t_0}^{t_1} dt \int\int_A \left\{ [\sigma_{zx}^+ - p_x^+]\left[-h\frac{\partial(\delta w)}{\partial x} \right] + [-\sigma_{zx}^- - p_x^-]\left[h\frac{\partial(\delta w)}{\partial x} \right] \right.$$

$$+ [\sigma_{zy}^+ - p_y^+]\left[-h\frac{\partial(\delta w)}{\partial y} \right] + [-\sigma_{zy}^- - p_y^-]\left[h\frac{\partial(\delta w)}{\partial y} \right]$$

$$\left. + [\sigma_{zz}^+ - p_z^+]\delta w + [-\sigma_{zz}^- - p_z^-]\delta w \right\} dx\,dy$$

$$- \int_{t_0}^{t_1} dt \oint_C \left\{ [M_x n_x + M_{xy} n_y - p_x^{(1)}]\left[-\frac{\partial(\delta w)}{\partial x} \right] \right.$$

$$+ [M_{yx} n_x + M_y n_y - p_y^{(1)}]\left[-\frac{\partial(\delta w)}{\partial y} \right]$$

$$\left. + [Q_x n_x + Q_y n_y - p_z^{(0)}]\delta w \right\} ds = 0, \tag{3.1.2}$$

where

$$f_x^{(1)} = \int_{-h}^h f_x z\,dz, \quad f_y^{(1)} = \int_{-h}^h f_y z\,dz, \quad f_z^{(0)} = \int_{-h}^h f_z\,dz$$

$$p_x^{(1)} = \int_{-h}^h p_x z\,dz, \quad p_y^{(1)} = \int_{-h}^h p_y z\,dz, \quad p_z^{(0)} = \int_{-h}^h p_z\,dz \tag{3.1.3}$$

are the plate body forces and plate surface tractions on the cylindrical boundary, and the superscripts (0) and (1) refer to quantities of the zeroth and first order, respectively, in the thickness coordinate. In addition, n_x and n_y are the direction cosines of an outward normal to the cylindrical boundary surface of the plate, and

$$(M_x, M_y, M_{xy}) = \int_{-h}^{h} (\sigma_{xx}, \sigma_{yy}, \sigma_{xy}) \, z \, dz$$

$$(Q_x, Q_y) = \int_{-h}^{h} (\sigma_{zx}, \sigma_{zy}) \, dz$$

(3.1.4)

are the plate stresses. These are the ordinary bending and twisting moments and transverse shearing forces.

The first integral in Eq. (3.1.2) is obtained by applying the volume integral in Eq. (1.4.9) to the entire volume of the plate. The next two integrals are obtained by applying the surface integral in Eq. (1.4.9) to the plane boundaries at $z = \pm h$ and the cylindrical boundary, respectively, under the assumption that only surface tractions, but not displacements, may be prescribed at $z = \pm h$. The two double integrals in the result are then combined so that the stresses $\sigma_{zx}^{\pm}$, $\sigma_{zy}^{\pm}$, and $\sigma_{zz}^{\pm}$ at $z = \pm h$ are eliminated; thus,

$$\int_{t_0}^{t_1} dt \int \int_A \left\{ \left[\frac{\partial M_x}{\partial x} + \frac{\partial M_{yz}}{\partial y} - Q_x + P_x^{(1)} \right. \right.$$

$$\left. + f_x^{(1)} + \rho \frac{2h^3}{3} \frac{\partial \ddot{w}}{\partial x} \right] \left[-\frac{\partial (\delta w)}{\partial x} \right]$$

$$+ \left[\frac{\partial M_{xy}}{\partial x} + \frac{\partial M_y}{\partial y} - Q_y + P_y^{(1)} \right.$$

$$\left. + f_y^{(1)} + \rho \frac{2h^3}{3} \frac{\partial \ddot{w}}{\partial y} \right] \left[-\frac{\partial (\delta w)}{\partial y} \right]$$

$$+ \left. \left[\frac{\partial Q_x}{\partial x} + \frac{\partial Q_y}{\partial y} + P_z^{(0)} + f_z^{(0)} - 2\rho h \ddot{w} \right] \delta w \right\} dx \, dy$$

$$- \int_{t_0}^{t_1} dt \oint_C \{\ldots\} \, ds = 0,$$

(3.1.5)

where

$$P_x^{(1)} = h(p_x^+ - p_x^-), \quad P_y^{(1)} = h(p_y^+ - p_y^-), \quad P_z^{(0)} = (p_z^+ + p_z^-) \quad (3.1.6)$$

are the plate surface tractions at $z = \pm h$.

By rearranging terms in the double integral, Eq. (3.1.5) first becomes

$$\int_{t_0}^{t_1} dt \int \int_A \left(\frac{\partial^2 M_x}{\partial x^2} + 2\frac{\partial^2 M_{xy}}{\partial x \partial y} + \frac{\partial^2 M_y}{\partial y^2} \right.$$

$$+ \frac{\partial P_x^{(1)}}{\partial x} + \frac{\partial P_y^{(1)}}{\partial y} + P_z^{(0)}$$

$$
+ \frac{\partial f_x^{(1)}}{\partial x} + \frac{\partial f_y^{(1)}}{\partial y} + f_z^{(0)}
$$

$$
- 2\rho h \ddot{w} + \rho \frac{2h^3}{3} \nabla^2 \ddot{w} \Bigg] \delta w
$$

$$
- \frac{\partial}{\partial x} \Bigg\{ \Bigg[\frac{\partial M_x}{\partial x} + \frac{\partial M_{yx}}{\partial y} - Q_x + P_x^{(1)}
$$

$$
+ f_x^{(1)} + \rho \frac{2h^3}{3} \frac{\partial \ddot{w}}{\partial x} \Bigg] \delta w \Bigg\}
$$

$$
- \frac{\partial}{\partial y} \Bigg\{ \Bigg[\frac{\partial M_{xy}}{\partial x} + \frac{\partial M_y}{\partial y} - Q_y + P_y^{(1)}
$$

$$
+ f^{(1)} + \rho \frac{2h^3}{3} \frac{\partial \ddot{w}}{\partial y} \Bigg] \delta w \Bigg\} \Bigg) \, dx \, dy
$$

$$
- \int_{t_0}^{t_1} dt \oint_C \{\dots\} \, ds = 0.
$$

By transforming the underlined terms in the double integral into a contour integral by Gauss' theorem, and by combining the contour integrals so that the Q_x and Q_y terms are canceled out, the result becomes

$$
\int_{t_0}^{t_1} dt \int \int_A \Bigg[\frac{\partial^2 M_x}{\partial x^2} + 2 \frac{\partial^2 M_{xy}}{\partial x \partial y} + \frac{\partial^2 M_y}{\partial y^2}
$$

$$
+ \frac{\partial P_x^{(1)}}{\partial x} + \frac{\partial P_y^{(1)}}{\partial y} + P_z^{(0)}
$$

$$
+ \frac{\partial f_x^{(1)}}{\partial x} + \frac{\partial f_y^{(1)}}{\partial y} + f_z^{(0)}
$$

$$
- 2\rho h \ddot{w} + \rho \frac{2h^3}{3} \nabla^2 \ddot{w} \Bigg] \delta w \, dx \, dy
$$

$$
- \int_{t_0}^{t_1} dt \oint_C \Bigg(\Bigg\{ \Bigg[\frac{\partial M_x}{\partial x} + \frac{\partial M_{yx}}{\partial y} + P_x^{(1)}
$$

$$
+ f_x^{(1)} + \rho \frac{2h^3}{3} \frac{\partial \ddot{w}}{\partial x} \Bigg] n_x
$$

$$
+ \Bigg[\frac{\partial M_{xy}}{\partial x} + \frac{\partial M_y}{\partial y} + P_y^{(1)}
$$

$$
+ f_y^{(1)} + \rho \frac{2h^3}{3} \frac{\partial \ddot{w}}{\partial y} \Bigg] n_y - p_z^{(0)} \Bigg\} \delta w
$$

$$
- [M_x n_x + M_{xy} n_y - p_x^{(1)}] \delta \left(\frac{\partial w}{\partial x} \right)
$$

$$
- [M_{yx} n_x + M_y n_y - p_y^{(1)}] \delta \left(\frac{\partial w}{\partial y} \right) \Bigg) \, ds = 0.
$$

The final step in our manipulation consists of transforming the contour integral from x,y-coordinates to the normal and tangential coordinates n,s. Thus,

$$\int_{t_0}^{t_1} dt \int \int_A [\ldots] \delta w \, dx \, dy$$

$$- \int_{t_0}^{t_1} dt \oint_C \left\{ \left[\frac{\partial M_n}{\partial n} + \frac{\partial M_{ns}}{\partial s} + P_n^{(1)} + f_n^{(1)} - p_z^{(0)} \right. \right.$$

$$\left. + \rho \frac{2h^3}{3} \frac{\partial \ddot{w}}{\partial n} \right] \delta w$$

$$\left. - [M_n - p_n^{(1)}] \delta \left(\frac{\partial w}{\partial n} \right) - [M_{ns} - p_s^{(1)}] \delta \left(\frac{\partial w}{\partial s} \right) \right\} ds = 0.$$

The last term in the contour integral is

$$\oint_C [M_{ns} - p_s^{(1)}] \frac{\partial (\delta w)}{\partial s} \, ds$$

$$= \oint_C \left\{ \frac{\partial}{\partial s}[(M_{ns} - p_s^{(1)}) \delta w] - \frac{\partial}{\partial s}(M_{ns} - p_s^{(1)}) \ \delta w \right\} ds$$

$$= [(M_{ns} - p_s^{(1)}) \delta w]_C - \oint_C \frac{\partial}{\partial s}(M_{ns} - p_s^{(1)}) \ \delta w \ ds$$

$$= 0 - \oint_C \left(\frac{\partial M_{ns}}{\partial s} - \frac{\partial p_s^{(1)}}{\partial s} \right) \delta w \ ds,$$

where the term $[\ldots]_C$ vanishes since the expression in the brackets is continuous around C. Thus, we finally find

$$\int_{t_0}^{t_1} dt \int \int_A \left[\frac{\partial^2 M_x}{\partial x^2} + .2 \frac{\partial^2 M_{xy}}{\partial x \partial y} + \frac{\partial^2 M_y}{\partial y^2} \right.$$

$$+ \frac{\partial P_x^{(1)}}{\partial x} + \frac{\partial P_y^{(1)}}{\partial y} + P_z^{(0)}$$

$$+ \frac{\partial f_x^{(1)}}{\partial x} + \frac{\partial f_y^{(1)}}{\partial y} + f_z^{(0)}$$

$$\left. - 2\rho h \ddot{w} + \rho \frac{2h^3}{3} \nabla^2 \ddot{w} \right] \delta w \, dx \, dy$$

$$- \int_{t_0}^{t_1} dt \oint_{C_p} \left\{ \left[\frac{\partial M_n}{\partial n} + 2 \frac{\partial M_{ns}}{\partial s} + P_n^{(1)} + f_n^{(1)} \right. \right.$$

$$\left. - p_z^{(0)} - \frac{\partial p_s^{(1)}}{\partial s} + \rho \frac{2h^3}{3} \frac{\partial \ddot{w}}{\partial n} \right] \delta w$$

$$\left. - [M_n - p_n^{(1)}] \delta \left(\frac{\partial w}{\partial n} \right) \right\} ds = 0. \qquad (3.1.7)$$

Equation (3.1.7) is the two-dimensional variational equation of motion of the plate. According to the rules of calculus of variations, this yields

$$\frac{\partial^2 M_x}{\partial x^2} + 2\frac{\partial^2 M_{xy}}{\partial x \partial y} + \frac{\partial^2 M_y}{\partial y^2} + \frac{\partial P_x^{(1)}}{\partial x} + \frac{\partial P_y^{(1)}}{\partial y} + P_z^{(0)}$$

$$+ \frac{\partial f_x^{(1)}}{\partial x} + \frac{\partial f_y^{(1)}}{\partial y} + f_z^{(0)} - 2\rho h \ddot{w} + \rho\frac{2h^3}{3}\nabla^2 \ddot{w} = 0 \quad (\text{in } A) \quad (3.1.8)$$

$$\frac{\partial M_n}{\partial n} + 2\frac{\partial M_{ns}}{\partial s} + P_n^{(1)} + f_n^{(1)} - p_z^{(0)} - \frac{\partial p_s^{(1)}}{\partial s}$$

$$+ \rho\frac{2h^3}{3}\frac{\partial \ddot{w}}{\partial n} = 0 \quad \text{or} \quad w = \text{prescribed}$$

$$(\text{on } C), \quad (3.1.9)$$

$$M_n - p_n^{(1)} = 0 \quad \text{or} \quad \frac{\partial w}{\partial n} = \text{prescribed}$$

which are the stress equation of motion and boundary conditions, respectively. The part of the boundary C on which tractions are prescribed can be designated by C_p. Since variations of displacements vanish on the part of C on which displacements are prescribed, C becomes C_p automatically, and the two notations become indistinguishable in Eq. (3.1.7).

According to Eqs. (3.1.9), the boundary conditions for a simply supported edge are taken as

$$w = 0 \quad \text{and} \quad M_n = 0,$$

and those for a clamped edge are taken as

$$w = 0 \quad \text{and} \quad \frac{\partial w}{\partial n} = 0.$$

We note that surface tractions in all three directions of x, y, z have been accommodated in the stress equation of motion. We also note that the effect of rotatory inertia is automatically included, although the effect of transverse shear deformation is not. More importantly, we have just demonstrated that, in deriving plate equations of the classical type, the volume and surface integrals in the variational equation (1.4.9) must be used *jointly*.

3.1.2 Transverse Shearing Forces

The above derivation shows that the transverse shearing forces have eventually disappeared altogether from the final form of the variational equation of motion. However, these forces were still present in Eq. (3.1.5), which is an intermediate form of the variational equation. By setting equal to 0 the coefficients of $-\partial(\delta w)/\partial x$ and $-\partial(\delta w)/\partial y$ in the field integral in this equation, there results

$$Q_x = \frac{\partial M_x}{\partial x} + \frac{\partial M_{yx}}{\partial y} + P_x^{(1)} + f_x^{(1)} + \rho\frac{2h^3}{3}\frac{\partial \ddot{w}}{\partial x}, \quad Q_y \cdots$$

When the last three terms are neglected in each of these two expressions, the results reduce to the usual relations between shearing forces and bending and twisting moments.

In terms of the coordinates n and s, the first of the two traction boundary conditions in Eqs. (3.1.9) has the form

$$Q_n + \frac{\partial M_{ns}}{\partial s} = p_z^{(0)} + \frac{\partial p_s^{(1)}}{\partial s}.$$

For a free edge for which $p_z^{(0)}$ and $p_s^{(1)}$ vanish, this becomes

$$Q_n + \frac{\partial M_{ns}}{\partial s} = 0.$$

The other boundary condition for a free edge is given by the second of Eqs. (3.1.9):

$$M_n = 0.$$

3.1.3   Displacements, Strains, and Stresses

The classical flexural theory of a thin plate is based on Kirchhoff's hypothesis, which states that a line across the thickness of a plate that is originally straight and normal to the middle plane of the plate remains straight and normal to the deflected middle surface and retains the same original length. The transverse shear and normal strains are thus zero, which are consistent with the assumed displacements in Eqs. (3.1.1). The other three strains are not zero, and are given by

$$e_{xx} = \frac{\partial u_x}{\partial x} = z\,k_x = -z\frac{\partial^2 w}{\partial x^2}$$

$$e_{yy} = \frac{\partial u_y}{\partial y} = z\,k_y = -z\frac{\partial^2 w}{\partial y^2} \qquad (3.1.10)$$

$$e_{xy} = \frac{1}{2}\left(\frac{\partial u_x}{\partial y} + \frac{\partial u_y}{\partial x}\right) = z\,k_{xy} = -z\frac{\partial^2 w}{\partial x\partial y},$$

where the curvatures and twist k_x, k_y, and k_{xy} are the plate strains.

From the three-dimensional strain–stress relations of an isotropic elastic solid, we single out the following for the nonzero linear strains:

$$e_{xx} = \frac{1}{E}(\sigma_{xx} - \nu\sigma_{yy} - \nu\sigma_{zz})$$

$$e_{yy} = \frac{1}{E}(\sigma_{yy} - \nu\sigma_{xx} - \nu\sigma_{zz})$$

$$e_{xy} = \frac{\sigma_{xy}}{2G} = \frac{(1+\nu)\sigma_{xy}}{E},$$

from which the stresses are given by

$$\sigma_{xx} = \frac{E(e_{xx} + \nu e_{yy})}{1 - \nu^2} + \frac{\sigma_{zz}\nu}{1 - \nu}$$

$$\sigma_{yy} = \frac{E(e_{yy} + \nu e_{xx})}{1 - \nu^2} + \frac{\sigma_{zz}\nu}{1 - \nu} \tag{3.1.11}$$

$$\sigma_{xy} = \frac{E e_{xy}}{1 + \nu}.$$

By substituting Eqs. (3.1.10) and (3.1.11) into the integrals in Eqs. (3.1.4) and ignoring the integration of the transverse normal stress σ_{zz}, the moments are

$$M_x = D(k_x + \nu k_y) = -D\left(\frac{\partial^2 w}{\partial x^2} + \nu\frac{\partial^2 w}{\partial y^2}\right)$$

$$M_y = D(k_y + \nu k_x) = -D\left(\frac{\partial^2 w}{\partial y^2} + \nu\frac{\partial^2 w}{\partial x^2}\right) \tag{3.1.12}$$

$$M_{xy} = D(1 - \nu)k_{xy} = -D(1 - \nu)\frac{\partial^2 w}{\partial x \partial y},$$

where

$$D = \frac{E}{1 - \nu^2}\frac{2h^3}{3}$$

is the flexural rigidity of the plate. Equations (3.1.12) are the usual plate stress–strain– displacement relations.

Because the transverse shear deformation has been neglected, the transverse shear stress–strain relations simply do not exist. This is equivalent to assuming the transverse shear rigidity of the plate equal to infinity.

3.1.4 Displacement Equation of Motion

The displacement equation of motion of a classical plate is obtained readily by substituting the plate stress–displacement relations from Eqs. (3.1.12) into the stress equation of motion (3.1.8); thus,

$$D\nabla^4 w + \frac{\partial P_x^{(1)}}{\partial x} + \frac{\partial P_y^{(1)}}{\partial y} + P_z^{(0)} + \frac{\partial f_x^{(1)}}{\partial x} + \frac{\partial f_y^{(1)}}{\partial y}$$

$$+ f_z^{(0)} - 2\rho h\ddot{w} + \rho\frac{2h^3}{3}\nabla^2\ddot{w} = 0 \qquad \text{(in } A\text{)}. \tag{3.1.13}$$

The usual form of this equation is obtained when the surface tractions in the x- and y-directions, the body forces, and the rotatory inertia are neglected:

$$D\nabla^4 w + P_z^{(0)} - 2\rho h\ddot{w} = 0 \qquad \text{(in } A\text{)}. \tag{3.1.13a}$$

3.1.5 Work and Energy

To extend the concept of the strain energy function from elasticity theory to plate theory, we start with the linearized form of Eq. (1.3.7):

$$\delta U_0 = \sigma_{xx}\delta e_{xx} + \sigma_{xy}\delta e_{xy} + \sigma_{xz}\delta e_{xz}$$

$$+\sigma_{yx}\delta e_{yx} + \sigma_{yy}\delta e_{yy} + \sigma_{yz}\delta e_{yz}$$
$$+\sigma_{zx}\delta e_{zx} + \sigma_{zy}\delta e_{zy} + \sigma_{zz}\delta e_{zz}.$$

Since the only nonvanishing strains are the three given by Eqs. (3.1.10), we have

$$\delta U_0 = \sigma_{xx}z\delta k_x + \sigma_{yy}z\delta k_y + 2\sigma_{xy}z\delta k_{xy}. \tag{3.1.14}$$

This is next integrated over the thickness of the plate by introducing the plate strain– energy function:

$$U_p = \int_{-h}^{h} U_0 \, dz.$$

Integration of the left-hand side of Eq. (3.1.14) then yields δU_p, and integration of the right-hand side of the equation transforms the stresses into the plate stresses according to Eqs. (3.1.4); these lead to the result

$$\delta U_p = M_x\delta k_x + M_y\delta k_y + 2M_{xy}\delta k_{xy}. \tag{3.1.15}$$

By means of Eqs. (3.1.12), this can be expressed in terms of plate strains and plate displacement as follows:

$$\delta U_p = \delta\left(\tfrac{1}{2}D\{(k_x + k_y)^2 - 2(1 - v)(k_x k_y - k_{xy}^2)\}\right)$$
$$= \delta\left(\tfrac{1}{2}D\left\{(\nabla^2 w)^2 - 2(1 - v)\left[\frac{\partial^2 w}{\partial x^2}\frac{\partial^2 w}{\partial y^2} - \left(\frac{\partial^2 w}{\partial x\partial y}\right)^2\right]\right\}\right), \tag{3.1.16}$$

from which U_p can be written immediately. We also have

$$\delta U_p = \frac{\partial U_p}{\partial k_x}\delta k_x + \frac{\partial U_p}{\partial k_y}\delta k_y + 2\frac{\partial U_p}{\partial k_{xy}}\delta k_{xy} \tag{3.1.17}$$

and, by comparing with Eq. (3.1.15),

$$M_x = \frac{\partial U_p}{\partial k_x}, \quad M_y = \frac{\partial U_p}{\partial k_y}, \quad M_{xy} = \frac{\partial U_p}{\partial k_{xy}}. \tag{3.1.18}$$

The plate stresses and strains now are related to each other through the plate strain–energy function in a similar manner as the three-dimensional stresses and strains in elasticity theory. The total strain energy of the plate finally is obtained by integrating U_p over the plan area A of the plate:

$$U = \int\int_A U_p \, dx\, dy. \tag{3.1.19}$$

The virtual work done by the external forces on the plate is formulated next. These include the body forces acting throughout the volume V, the surface tractions on the plan area A of the boundary planes at $z = \pm h$, and the surface tractions on the cylindrical boundary surface S_p (Figure 3.1.1). Corresponding to the virtual displacement δw, the virtual work is thus

$$
\begin{aligned}
\delta W = \int \int_A & \left[-f_x^{(1)} \frac{\partial(\delta w)}{\partial x} - f_y^{(1)} \frac{\partial(\delta w)}{\partial y} + f_z^{(0)} \delta w \right] dx\, dy \\
+ \int_A \int & \left[-P_x^{(1)} \frac{\partial(\delta w)}{\partial x} - P_y^{(1)} \frac{\partial(\delta w)}{\partial y} + P_z^{(0)} \delta w \right] dx\, dy \\
+ \oint_{C_p} & \left[-p_x^{(1)} \frac{\partial(\delta w)}{\partial x} - p_y^{(1)} \frac{\partial(\delta w)}{\partial y} + p_z^{(0)} \delta w \right] ds.
\end{aligned} \qquad (3.1.20)
$$

The kinetic energy function

$$
T_0 = \frac{1}{2} \rho (\dot{u}_x^2 + \dot{u}_y^2 + \dot{u}_z^2)
$$

is the kinetic energy per unit volume of a solid. Upon integration, this yields the plate kinetic energy function

$$
T_p = \int_{-h}^{h} T_0 \, dz = \frac{1}{3} \rho h^3 \left[\left(\frac{\partial \dot{w}}{\partial x} \right)^2 + \left(\frac{\partial \dot{w}}{\partial y} \right)^2 \right] + \rho h \dot{w}^2,
$$

which is the kinetic energy per unit area of the plate. The total kinetic energy of the plate is then

$$
T = \int \int_A T_p \, dx\, dy. \qquad (3.1.21)
$$

Alternatively, the plate equations may be derived from Hamilton's principle by means of Eqs. (3.1.19) through (3.1.21). However, this will not be pursued here.

3.2   Refined Equations for Flexure of an Isotropic Plate: Mindlin Plate Equations and Timoshenko Beam Equations

The concept of a refined flexural theory was originally due to Timoshenko (1921), who introduced his famous beam equations by including the effects of transverse shear and rotatory inertia. The transverse shear effect was later introduced into the plate equations in the static case by Reissner (1945). This, together with the rotatory inertia, was then incorporated into the equations of motion of an isotropic plate by Mindlin (1951a). We shall now derive Mindlin's plate equations from the variational equation of motion in linear elasticity.

3.2.1   Variational Equation of Motion

The same plate in Figure 3.1.1 is considered, but the three-dimensional displacements are now taken in the form

$$u_x(x, y, z, t) = z\psi(x, y, t), \quad u_y(x, y, z, t) = z\phi(x, y, t)$$
$$u_z(x, y, z, t) = w(x, y, t), \tag{3.2.1}$$

where w, as before, is the deflection at the middle plane of the plate, and ψ and ϕ are the angles of rotation of a normal to the middle plane in the x- and y-directions, respectively.

Substituting Eqs. (3.2.1) into the linear variational equation (1.4.9) and carrying out integration with respect to z over the thickness of the plate, we find

$$\int_{t_0}^{t_1} dt \int \int_A \left\{ \left[\frac{\partial M_x}{\partial x} + \frac{\partial M_{yx}}{\partial y} - Q_x + h(\sigma_{zx}^+ + \sigma_{zx}^-) \right. \right.$$
$$\left. + f_x^{(1)} - \rho \frac{2h^3}{3} \ddot{\psi} \right] \delta\psi$$
$$+ \left[\frac{\partial M_{xy}}{\partial x} + \frac{\partial M_y}{\partial y} - Q_y + h(\sigma_{zy}^+ + \sigma_{zy}^-) \right.$$
$$\left. + f_y^{(1)} - \rho \frac{2h^3}{3} \ddot{\phi} \right] \delta\phi$$
$$+ \left[\frac{\partial Q_x}{\partial x} + \frac{\partial Q_y}{\partial y} + (\sigma_{zz}^+ - \sigma_{zz}^-) \right.$$
$$\left. \left. + f_z^{(0)} - 2\rho h \ddot{w} \right] \delta w \right\} dx\,dy$$
$$- \int_{t_0}^{t_1} dt \int \int_A \left\{ [\sigma_{zx}^+ - p_x^+](-h\delta\psi) + [-\sigma_{zx}^- - p_x^-](h\delta\psi) \right.$$
$$+ [\sigma_{zy}^+ - p_y^+](-h\delta\phi) + [-\sigma_{zy}^- - p_y^-](h\delta\phi)$$
$$\left. + [\sigma_{zz}^+ - p_z^+]\delta w + [-\sigma_{zz}^- - p_z^-]\delta w \right\} dx\,dy$$
$$- \int_{t_0}^{t_1} dt \oint_C \left\{ [M_x n_x + M_{xy} n_y - p_x^{(1)}]\delta\psi \right.$$
$$+ [M_{yx} n_x + M_y n_y - p_y^{(1)}]\delta\phi$$
$$\left. + [Q_x n_x + Q_y n_y - p_z^{(0)}]\delta w \right\} ds = 0. \tag{3.2.2}$$

By combining the first two integrals to eliminate $\sigma_{zx}^\pm$, $\sigma_{zy}^\pm$, and $\sigma_{zz}^\pm$, and transforming the third integral from x,y- to n,s-coordinates, there results

$$\int_{t_0}^{t_1} dt \int \int_A \left\{ \left[\frac{\partial M_x}{\partial x} + \frac{\partial M_{yx}}{\partial y} - Q_x + P_x^{(1)} + f_x^{(1)} - \frac{2}{3}\rho h^3 \ddot{\psi} \right] \delta\psi \right.$$
$$+ \left[\frac{\partial M_{xy}}{\partial x} + \frac{\partial M_y}{\partial y} - Q_y + P_y^{(1)} + f_y^{(1)} - \frac{2}{3}\rho h^3 \ddot{\phi} \right] \delta\phi$$

$$+ \left[\frac{\partial Q_x}{\partial x} + \frac{\partial Q_y}{\partial y} + P_z^{(0)} + f_z^{(0)} - 2\rho h \ddot{w} \right] \delta w \Bigg\} \, dx \, dy$$

$$- \int_{t_0}^{t_1} dt \oint_C \{ [M_n - p_n^{(1)}] \delta \psi_n + [M_{ns} - p_s^{(1)}] \delta \psi_s + [Q_n - p_z^{(0)}] \delta w \} \, ds = 0.$$

$$(3.2.3)$$

This is the two-dimensional variational equation of motion in the refined plate theory, from which the stress equations of motion are

$$\frac{\partial M_x}{\partial x} + \frac{\partial M_{yx}}{\partial y} - Q_x + P_x^{(1)} + f_x^{(1)} - \frac{2}{3}\rho h^3 \ddot{\psi} = 0$$

$$\frac{\partial M_{xy}}{\partial x} + \frac{\partial M_y}{\partial y} - Q_y + P_y^{(1)} + f_y^{(1)} - \frac{2}{3}\rho h^3 \ddot{\phi} = 0 \qquad \text{(in } A\text{)} \quad (3.2.4)$$

$$\frac{\partial Q_x}{\partial x} + \frac{\partial Q_y}{\partial y} + P_z^{(0)} + f_z^{(0)} - 2\rho h \ddot{w} = 0,$$

and the boundary conditions are

$$M_n = p_n^{(1)} \quad \text{or} \quad \psi_n = \text{prescribed}$$

$$M_{ns} = p_s^{(1)} \quad \text{or} \quad \psi_s = \text{prescribed} \qquad \text{(on } C\text{)} \qquad (3.2.5)$$

$$Q_n = p_z^{(0)} \quad \text{or} \quad w = \text{prescribed}.$$

As was noted first by Reissner and then by Mindlin, the twisting moment M_{ns} and shearing force Q_n may now be prescribed independently. For a free edge, the boundary conditions are taken as

$$M_n = M_{ns} = Q_n = 0.$$

We also note that, because the plate displacements assumed in Eqs. (3.2.1) are uncoupled, the stress equations of motion and boundary conditions are derived separately from the volume and surface integrals in the three-dimensional variational equation. This makes the derivation considerably simpler than that of the classical plate equations given in the preceding section.

3.2.2 Displacements, Strains, and Stresses

According to the displacements assumed in Eqs. (3.2.1), the engineering transverse shear strains in the plate are given by

$$\gamma_{zx} = \psi + \frac{\partial w}{\partial x}, \qquad \gamma_{zy} = \phi + \frac{\partial w}{\partial y}.$$

To suppress the transverse shear strains, we simply let

$$\psi = -\frac{\partial w}{\partial x}, \qquad \phi = -\frac{\partial w}{\partial y},$$

and the classical case is retrieved.

Substitution of Eqs. (3.2.1) into (1.1.6) yields the linear strains

$$e_{xx} = z k_x = z \frac{\partial \psi}{\partial x}, \qquad e_{yy} = z k_y = z \frac{\partial \phi}{\partial y}$$

$$e_{xy} = z k_{xy} = \frac{1}{2} z \left(\frac{\partial \psi}{\partial y} + \frac{\partial \phi}{\partial x} \right) \tag{3.2.6}$$

$$e_{zx} = k_{zx} = \frac{1}{2} \left(\psi + \frac{\partial w}{\partial x} \right), \qquad e_{zy} = k_{zy} = \frac{1}{2} \left(\phi + \frac{\partial w}{\partial y} \right), \qquad e_{zz} = 0,$$

where k_x, k_y, k_{xy}, k_{zx}, and k_{zy} are the plate strains. Relations between M_x, M_y, M_{xy} and k_x, k_y, k_{xy} remain the same as in the classical case, from which we find the following plate stress–strain–displacement relations for the refined case:

$$M_x = D(k_x + \nu k_y) = D \left(\frac{\partial \psi}{\partial x} + \nu \frac{\partial \phi}{\partial y} \right)$$

$$M_y = D(k_y + \nu k_x) = D \left(\frac{\partial \phi}{\partial y} + \nu \frac{\partial \psi}{\partial x} \right) \tag{3.2.7}$$

$$M_{xy} = D(1 - \nu) k_{xy} = \frac{1}{2} D(1 - \nu) \left(\frac{\partial \psi}{\partial y} + \frac{\partial \phi}{\partial x} \right).$$

Since the transverse shear deformations no longer vanish, the transverse shear stress–strain relations must now be considered. In linear elasticity, these have the form

$$\sigma_{zx} = 2G e_{zx}, \qquad \sigma_{zy} = 2G e_{zy}, \tag{3.2.8}$$

where e_{zx} and e_{zy} still are given by Eqs. (3.2.6). Substitution of Eqs. (3.2.8) into (3.1.4) yields the plate transverse shearing forces

$$Q_x = 4\kappa G h\, k_{zx} = 2\kappa G h \left(\psi + \frac{\partial w}{\partial x} \right)$$

$$Q_y = 4\kappa G h\, k_{zy} = 2\kappa G h \left(\phi + \frac{\partial w}{\partial y} \right), \tag{3.2.9}$$

where κ is a shear factor introduced in the manner of Mindlin (1951a). As will be shown later in this section, the shear factor is equal to $\pi^2/12$. The value is determined differently in the static theory given by Reissner (1945).

3.2.3 Work and Energy

The discussion on work and energy in Section 3.1 may be readily extended by further including the contributions of the transverse shear stresses and strains. Thus, instead of Eqs. (3.1.14), (3.1.15), and (3.1.16), we now have

$$\delta U_0 = \sigma_{xx} z\, \delta k_x + \sigma_{yy} z\, \delta k_y + 2\sigma_{xy} z\, \delta k_{xy} + 2\sigma_{zx}\, \delta k_{zx} + 2\sigma_{zy}\, \delta k_{zy}$$

$$\delta U_P = M_x\,\delta k_x + M_y\,\delta k_y + 2M_{xy}\delta k_{xy} + 2Q_x\,\delta k_{zx} + 2Q_y\,\delta k_{zy}$$

$$= \delta\left(\frac{1}{2}D\{(k_x + k_y)^2 - 2(1-v)(k_x k_y - k_{xy}^2)\}\right.$$

$$\left. + 4\kappa Gh(k_{zx}^2 + k_{zy}^2)\right)$$

$$= \delta\left(\frac{1}{2}D\left\{(\nabla^2 w)^2 - 2(1-v)\left[\frac{\partial^2 w}{\partial x^2}\frac{\partial^2 w}{\partial y^2} - \frac{\partial^2 w}{\partial x\partial y}^2\right]\right\}\right.$$

$$\left. + \kappa Gh\left[\left(\psi + \frac{\partial w}{\partial x}\right)^2 + \left(\phi + \frac{\partial w}{\partial y}\right)^2\right]\right).$$

From these, we also have

$$\delta U_p = \frac{\partial U_p}{\partial k_x}\delta k_x + \frac{\partial U_p}{\partial k_y}\delta k_y + 2\frac{\partial U_p}{\partial k_{xy}}\delta k_{xy}$$

$$+ 2\frac{\partial U_p}{\partial k_{zx}}\delta k_{zx} + 2\frac{\partial U_p}{\partial k_{zy}}\delta k_{zy}$$

and

$$M_x = \frac{\partial U_p}{\partial k_x}, \quad M_y = \frac{\partial U_p}{\partial k_y}, \quad M_{xy} = \frac{\partial U_p}{\partial k_{xy}},$$

$$Q_x = \frac{\partial U_p}{\partial k_{zx}}, \quad Q_y = \frac{\partial U_p}{\partial k_{zy}}.$$

The plate strain energy function U_p can be written immediately from δU_p, and the total strain energy U of the plate is then

$$U = \int\int_A U_p\,dx\,dy$$

$$= \int\int_A \left[\frac{1}{2}D\left\{(\nabla^2 w)^2 - 2(1-v)\left(\frac{\partial^2 w}{\partial x^2}\frac{\partial^2 w}{\partial y^2} - \frac{\partial^2 w}{\partial x\partial y}^2\right)\right\}\right.$$

$$\left. + \kappa Gh\left\{\left(\psi + \frac{\partial w}{\partial x}\right)^2 + \left(\phi + \frac{\partial w}{\partial y}\right)^2\right\}\right]dx\,dy.$$

In a similar manner, the virtual work and kinetic energy may be obtained from Eqs. (3.1.20) and (3.1.21) by replacing $-\partial w/\partial x$ and $-\partial w/\partial y$ by ψ and ϕ, respectively. The results are

$$\delta W = \int\int_A [f_x^{(1)}\,\delta\psi + f_y^{(1)}\,\delta\phi + f_z^{(0)}\,\delta w]\,dx\,dy$$

$$\int\int_A [P_x^{(1)}\,\delta\psi + P_y^{(1)}\,\delta\phi + P_z^{(0)}\,\delta w]\,dx\,dy$$

$$T = \int \int_A \left[\frac{1}{3}\rho h^3 (\psi^2 + \phi^2) + \rho h \dot{w}^2 \right] dx\, dy.$$

3.2.4   Displacement Equations of Motion

By substituting the plate stresses from Eqs. (3.2.7) and (3.2.9) into the stress equations (3.2.4), the following displacement equations of motion are obtained:

$$\frac{1}{2}D(1-v)\nabla^2\psi + \frac{1}{2}D(1+v)\frac{\partial}{\partial x}\left(\frac{\partial\psi}{\partial x} + \frac{\partial\phi}{\partial y}\right)$$

$$- 2\kappa Gh\left(\psi + \frac{\partial w}{\partial x}\right) + P_x^{(1)} + f_x^{(1)} - \rho\frac{2h^3}{3}\ddot{\psi} = 0$$

$$\frac{1}{2}D(1-v)\nabla^2\phi + \frac{1}{2}D(1+v)\frac{\partial}{\partial y}\left(\frac{\partial\psi}{\partial x} + \frac{\partial\phi}{\partial y}\right) \qquad \text{(in } A\text{)} \quad (3.2.10)$$

$$- 2\kappa Gh\left(\phi + \frac{\partial w}{\partial y}\right) + P_y^{(1)} + f_y^{(1)} - \rho\frac{2h^3}{3}\ddot{\phi} = 0$$

$$2\kappa Gh\left(\nabla^2 w + \frac{\partial\psi}{\partial x} + \frac{\partial\phi}{\partial y}\right) + P_z^{(0)} + f_z^{(0)} - 2\rho h\ddot{w} = 0.$$

3.2.5   Timoshenko Beam Equations

To reduce Eqs. (3.2.10) to the equations of a beam, we suppress the dependence on y and replace the plate quantities by the corresponding beam quantities. Specifically, D, $2h$, $2h^3/3$, and $P_z^{(0)}$ are replaced by the beam bending rigidity EI, cross-sectional area A, moment of inertia I, and lateral load intensity q, respectively. The results are

$$EI\frac{\partial^2\psi}{\partial x^2} - \kappa GA\left(\psi + \frac{\partial w}{\partial x}\right) - \rho I\ddot{\psi} = 0$$

$$\kappa GA\frac{\partial}{\partial x}\left(\psi + \frac{\partial w}{\partial x}\right) + q - \rho A\ddot{w} = 0, \qquad (3.2.11)$$

which are the well-known Timoshenko beam equations (Timoshenko 1921).

3.2.6   Determination of Shear Factor

Mindlin (1951a) proposed to determine the shear factor by matching the frequency of the lowest simple thickness-shear mode of vibration of an infinite plate given by the refined plate equations with that given by the exact elasticity theory. From the exact elasticity theory, this frequency was found to be given by Eq. (2.3.7). From the plate equations (3.2.10), we ignore the surface and body forces and let

$$\psi = \Psi\, e^{i\omega t}, \quad w = \phi = 0.$$

The first of these equations then yields

$$2\kappa Gh - \rho\frac{2h^3}{3}\omega^2 = 0.$$

By solving for ω and equating the result to the lowest frequency given by Eq. (2.3.7), with $q = 1$, the shear factor is found to be

$$\kappa = \frac{\pi^2}{12}.$$

3.3   Classical Equations for Extension of an Isotropic Plate

In this section, the classical equations for extension of an isotropic plate are derived. The method of derivation is similar to that used for flexure in the preceding sections, through the use of the linear variational equation of motion. The concepts of work and energy were covered for plate flexure, but will not be covered for plate extension.

3.3.1   Variational Equation of Motion

Consider again the plate in Figure 3.1.1. In the classical theory of extension of a thin plate, the three-dimensional displacements are taken in the form

$$u_x(x, y, z, t) = u(x, y, t), \quad u_y(x, y, z, t) = v(x, y, t), \quad u_z(x, y, z, t) = 0,$$

$$(3.3.1)$$

where u and v are the two-dimensional plate displacements. By substituting Eqs. (3.3.1) into the linear variational equation (1.4.9) and carrying out integration with respect to z over the thickness of the plate as before, we find

$$\int_{t_0}^{t_1} dt \int\int_A \left\{ \left[\frac{\partial N_x}{\partial x} + \frac{\partial N_{xy}}{\partial y} + \sigma_{zx}^+ - \sigma_{zx}^- + f_x^{(0)} - 2\rho h\ddot{u}\right] \delta u \right.$$

$$\left. + \left[\frac{\partial N_{xy}}{\partial x} + \frac{\partial N_y}{\partial y} + \sigma_{zy}^+ - \sigma_{zy}^- + f_y^{(0)} - 2\rho h\ddot{v}\right] \delta v \right\} dx\, dy$$

$$- \int_{t_0}^{t_1} dt \int_A \int \left\{ [(\sigma_{zx}^+ - p_x^+) + (-\sigma_{zx}^- - p_x^-)]\delta u \right.$$

$$\left. + [(\sigma_{zy}^+ - p_y^+) + (-\sigma_{zy}^- - p_y^-)]\delta v \right\} dx\, dy$$

$$- \int_{t_0}^{t_1} dt \oint_C \left\{ [N_x n_x + N_{xy} n_y - p_x^{(0)}]\delta u \right.$$

$$\left. + [N_{xy} n_x + N_y n_y - p_y^{(0)}]\delta v \right\} ds = 0.$$

By further combining the two double integrals and transforming from rectangular to normal and tangential coordinates in the contour integral, this becomes

$$\int_{t_0}^{t_1} dt \int\int_A \left\{ \left[\frac{\partial N_x}{\partial x} + \frac{\partial N_{xy}}{\partial y} + P_x^{(0)} + f_x^{(0)} - 2\rho h\ddot{u}\right] \delta u \right.$$

$$+ \left[\frac{\partial N_{xy}}{\partial x} + \frac{\partial N_y}{\partial y} + P_y^{(0)} + f_y^{(0)} - 2\rho h \ddot{v} \right] \delta v \right\} dx\, dy$$

$$- \int_{t_0}^{t_1} dt \oint_C \{ [N_n - p_n^{(0)}] \delta u_n + [N_{ns} - p_s^{(0)}] \delta u_s \}\, ds = 0. \quad (3.3.2)$$

In Eq. (3.3.2), we have, similar to Eqs. (3.1.3) and (3.1.6), the plate tractions and plate body forces

$$(p_x^{(0)}, p_y^{(0)}, p_n^{(0)}, p_s^{(0)}) = \int_{-h}^{h} (p_x, p_y, p_n, p_s)\, dz$$

$$P_x^{(0)} = p_x^+ + p_x^-, \qquad P_y^{(0)} = p_y^+ + p_y^- \qquad (3.3.3)$$

$$f_x^{(0)} = \int_{-h}^{h} f_x\, dz, \qquad f_y^{(0)} = \int_{-h}^{h} f_y\, dz.$$

The plate stresses are now the in-plane forces

$$(N_x, N_y, N_{xy}) = \int_{-h}^{h} (\sigma_{xx}, \sigma_{yy}, \sigma_{xy})\, dz. \qquad (3.3.4)$$

Equation (3.3.2) is the two-dimensional variational equation of motion for plate extension, from which we obtain the stress equations of motion

$$\frac{\partial N_x}{\partial x} + \frac{\partial N_{xy}}{\partial y} + P_z^{(0)} + f_x^{(0)} - 2\rho h \ddot{u} = 0$$

$$\frac{\partial N_{xy}}{\partial x} + \frac{\partial N_y}{\partial y} + P_y^{(0)} + f_y^{(0)} - 2\rho h \ddot{v} = 0$$

$$\text{(in } A) \qquad (3.3.5)$$

and the boundary conditions

$$N_n - p_n^{(0)} = 0 \quad \text{or} \quad u_n = \text{prescribed}$$

$$N_{ns} - p_s^{(0)} = 0 \quad \text{or} \quad u_s = \text{prescribed}.$$

$$\text{(on } C) \qquad (3.3.6)$$

3.3.2 *Displacements, Strains, and Stresses*

The strain–displacement relations are simply

$$e_{xx} = \frac{\partial u}{\partial x}, \quad e_{yy} = \frac{\partial v}{\partial y}, \quad e_{xy} = \frac{1}{2}\left(\frac{\partial u}{\partial y} + \frac{\partial v}{\partial x} \right). \qquad (3.3.7)$$

To derive the plate stress–strain relations, we again make use of the three-dimensional relations in Eqs. (3.1.11). Substituting these in Eqs. (3.3.4) and neglecting the contributions of σ_{zz} in the integration as before, we find

$$N_x = C(e_{xx} + \nu e_{yy})$$
$$N_y = C(e_{yy} + \nu e_{xx})$$
$$N_{xy} = C(1 - \nu)e_{xy},$$

$$(3.3.8)$$

where

$$C = \frac{2Eh}{1 - \nu^2}$$

is the extensional rigidity of the plate. By substituting Eqs. (3.3.7) into Eqs. (3.3.8), the plate stress– displacement relations are obtained.

3.3.3   Displacement Equations of Motion

The displacement equations of motion are finally obtained by substituting the plate stress– displacement relations into the stress equations (3.3.5). The results are

$$\frac{Eh}{1+\nu}\nabla^2 u + \frac{Eh}{1-\nu}\frac{\partial}{\partial x}\left(\frac{\partial u}{\partial x} + \frac{\partial v}{\partial y}\right)$$
$$+ P_x^{(0)} + f_x^{(0)} - 2\rho\ddot{u} = 0$$
$$\frac{Eh}{1+\nu}\nabla^2 v + \frac{Eh}{1-\nu}\frac{\partial}{\partial y}\left(\frac{\partial u}{\partial x} + \frac{\partial v}{\partial y}\right)$$
$$+ P_y^{(0)} + f_y^{(0)} - 2\rho\ddot{v} = 0. \tag{3.3.9}$$

3.4   Refined Equations for Extension of an Isotropic Plate

A refined linear theory for extension of an isotropic plate was developed by Kane and Mindlin (1956) by including the thickness-stretch effect. However, the theory is not dependable for a Poisson ratio greater than 1/3, as will be seen presently.

3.4.1   Variational Equation of Motion

The three-dimensional displacements are taken in the form

$$u_x(x, y, z, t) = u(x, y, t), \quad u_y(x, y, z, t) = v(x, y, t)$$
$$u_z(x, y, z, t) = z\beta(x, y, t), \tag{3.4.1}$$

where u and v are the same as in the previous section and β is an additional plate displacement, accommodating the thickness-stretch deformation in the plate.

By substituting the displacements from Eqs. (3.4.1) into the linear variational equation (1.4.9) and carrying out integration over the thickness of the plate in a similar manner as before, there results

$$\int_{t_0}^{t_1} dt \int_A \int \left\{ \left[\frac{\partial N_x}{\partial x} + \frac{\partial N_{xy}}{\partial y} + P_x^{(0)} + f_x^{(0)} - 2\rho h \ddot{u} \right] \delta u \right.$$
$$+ \left[\frac{\partial N_{xy}}{\partial x} + \frac{\partial N_y}{\partial y} + P_y^{(0)} + f_y^{(0)} - 2\rho h \ddot{v} \right] \delta v$$
$$+ \left[\frac{\partial S_x}{\partial x} + \frac{\partial S_y}{\partial y} - Q_z + P_z^{(1)} + f_z^{(1)} \right.$$

$$-\frac{2\rho h^3}{3}\ddot{\beta}\Big]\,\delta\beta\Big\}\,dx\,dy$$

$$-\int_{t_0}^{t_1} dt \oint_C \{[N_n - p_n^{(0)}]\,\delta u_n + [N_{ns} - p_s^{(0)}]\,\delta u_s$$

$$+\ [S_n - p_n^{(1)}]\,\delta\beta\}\,ds = 0, \tag{3.4.2}$$

where we have, in addition to Eqs. (3.3.3) and (3.3.4),

$$P_z^{(1)} = h(p_z^+ - p_z^-)$$

$$f_z^{(1)} = \int_{-h}^{h} f_z\, z\, dz, \qquad p_n^{(1)} = \int_{-h}^{h} p_n\, z\, dz \tag{3.4.3}$$

$$(S_x, S_y, Q_z) = \int_{-h}^{h} (\sigma_{zx} z, \sigma_{zy} z, \sigma_{zz})\, dz. \tag{3.4.4}$$

Among the additional plate stresses in Eqs. (3.4.4), Q_z is equal to the average transverse normal stress multiplied by the plate thickness, and S_x and S_y are the components of pinching shear, a term used by Kane and Mindlin. The latter play a role in extension similar to the transverse shearing forces in flexure. Equation (3.4.2) is the two-dimensional variational equation of motion for the plate, from which the stress equations of motion and boundary conditions may be written in the usual manner.

3.4.2 Displacements, Strains, and Stresses

The strain–displacement relations are

$$e_{xx} = \frac{\partial u}{\partial x}, \quad e_{yy} = \frac{\partial v}{\partial y}, \quad e_{xy} = \frac{1}{2}\left(\frac{\partial u}{\partial y} + \frac{\partial v}{\partial x}\right)$$

$$e_{zz} = \beta, \quad e_{zx} = \frac{1}{2}z\frac{\partial\beta}{\partial x}, \quad e_{zy} = \frac{1}{2}z\frac{\partial\beta}{\partial y}, \tag{3.4.5}$$

and the stress–strain relations are

$$\sigma_{xx} = (\lambda + 2\mu)\,e_{xx} + \lambda(e_{yy} + \kappa' z_z)$$

$$\sigma_{yy} = (\lambda + 2\mu)\,e_{yy} + \lambda(e_{xx} + \kappa' e_{zz})$$

$$\sigma_{zz} = (\lambda + 2\mu)\kappa'^2\, e_{zz} + \lambda\kappa'(e_{xx} + e_{yy}) \tag{3.4.6}$$

$$\sigma_{yz} = 2\mu\, e_{yz}, \quad \sigma_{zx} = 2\mu\, e_{zx}, \quad \sigma_{xy} = 2\mu\, e_{xy},$$

where κ' is a factor similar to the shear factor used by Mindlin in his flexural equations. Substitution of Eqs. (3.4.5) into (3.4.6) and the results in turn into Eqs. (3.3.4) and (3.4.4) yields

$$N_x = 2h\left[(\lambda + 2\mu)\frac{\partial u}{\partial x} + \lambda\left(\frac{\partial v}{\partial y} + \kappa'\beta\right)\right]$$

$$N_y = 2h\left[(\lambda + 2\mu)\frac{\partial v}{\partial y} + \lambda\left(\frac{\partial u}{\partial x} + \kappa'\beta\right)\right]$$

$$N_{xy} = 2\mu h\left(\frac{\partial u}{\partial y} + \frac{\partial v}{\partial x}\right)$$

$$S_x = \frac{2\mu h^3}{3}\frac{\partial \beta}{\partial x}$$

$$S_y = \frac{2\mu h^3}{3}\frac{\partial \beta}{\partial y}$$

$$Q_z = 2h\left[\kappa'^2(\lambda + 2\mu)\beta + \kappa\lambda\left(\frac{\partial u}{\partial x} + \frac{\partial v}{\partial y}\right)\right],$$

$$(3.4.7)$$

which are the plate stress–displacement relations. To reduce these to those given earlier in Section 3.3, the results for S_x and S_y are merely dropped, and the result for Q_z is used to eliminate β. Thus, by setting $Q_z = 0$, solving for β and substituting the result into N_x and N_y, the earlier results in Eqs. (3.3.8) are obtained.

3.4.3 Displacement Equations of Motion

Substitution of Eqs. (3.4.7) into the stress equations of motion yields the following displacement equations of motion:

$$2\mu h\,\nabla^2 u + 2(\lambda + \mu)h\frac{\partial}{\partial x}\left(\frac{\partial u}{\partial x} + \frac{\partial v}{\partial y}\right) + 2\kappa'\lambda h\frac{\partial \beta}{\partial x}$$

$$+ P_x^{(0)} + f_x^{(0)} - 2\rho h\ddot{u} = 0$$

$$2\mu h\,\nabla^2 v + 2(\lambda + \mu)h\frac{\partial}{\partial y}\left(\frac{\partial u}{\partial x} + \frac{\partial v}{\partial y}\right) + 2\kappa'\lambda h\frac{\partial \beta}{\partial y}$$

$$+ P_y^{(0)} + f_y^{(0)} - 2\rho h v = 0$$

$$(3.4.8)$$

$$\frac{2\mu h^3}{3}\nabla^2\beta - 2\kappa'^2(\lambda + 2\mu)h\beta - 2\kappa'\lambda h\left(\frac{\partial u}{\partial x} + \frac{\partial v}{\partial y}\right)$$

$$+ P_z^{(1)} + f_z^{(1)} - \tfrac{2}{3}\rho h^3\ddot{\beta} = 0.$$

The factor κ' can be determined in a similar manner as the shear factor in the flexural case, by matching the frequency of simple thickness-stretch vibration given by these plate equations with that given by the exact elasticity theory. Thus, Kane and Mindlin (1956) found

$$\kappa' = \frac{\pi^2}{12},$$

which has the same value as the shear factor used in plate flexure by Mindlin (1951a).

3.4.4   Higher Order Theory of Plate Extension

The above theory of plate extension accommodates only the thickness deformation that is associated with the lowest symmetric thickness-stretch mode, the frequency of which increases with increasing Poisson's ratio v. For $v > 1/3$, the frequency of this mode becomes higher than the frequency of the lowest symmetric thickness-shear mode, which is independent of v according to the results of the exact elasticity theory. Since the latter mode is not included in the above plate theory, frequencies predicated for modes higher than the lowest may become unreliable.

To improve on the above theory of plate extension, the lowest symmetric thickness-shear mode of the plate must be further included. This was later accomplished by Mindlin and Medick (1959) in the construction of a still higher order theory of plate extension. Their theory is valid for a much higher frequency range and will not be covered here.

3.5   Vibrations of an Infinite Plate: Useful Ranges of Plate Equations

For the plate equations that have been derived in the preceding sections, it will be necessary to know their useful ranges of applicability in terms of frequency and wavelength. These are determined by applying the plate equations to the analysis of free vibration in an infinite isotropic plate and by comparing the results with those given by the exact elasticity equations. An important consideration is to decide whether classical or refined plate equations should be used in an engineering analysis. Generally speaking, classical equations are simpler to apply, but refined equations can be more accurate. Among other things, the choice often depends on the frequency range involved in the analysis.

3.5.1   Flexural Vibration Based on Classical Equations

In the classical theory of plate flexure, the displacement equation of motion is given by Eq. (3.1.13). For free vibration, surface tractions and body forces are absent. For plane–strain modes in an infinite isotropic plate, the deflection is taken in the form

$$w = W \sin \xi x \, e^{i\omega t}, \tag{3.5.1}$$

which depends on x only. Equation (3.1.13) then yields the frequency equation

$$(1 + \kappa_r \xi^2 h^2/3)\omega^2 = \frac{D}{2\rho h}\xi^4 \quad \text{or} \quad \omega = \xi^2 h \sqrt{\frac{E}{3\rho(1 - v^2)(1 + \kappa_r \xi^2 h^2/3)}} \tag{3.5.2}$$

where the κ_r term reflects the effect of rotatory inertia. By dropping this term, rotatory inertia is eliminated, and the result reduces to

$$\omega^2 = \frac{D}{2\rho h}\xi^4 \quad \text{or} \quad \omega = \xi^2 h\sqrt{\frac{E}{3\rho(1-\nu^2)}}, \tag{3.5.3}$$

which also may be derived directly from Eq. (3.1.13a). Indeed, Eq. (3.5.2) reduces to (3.5.3) as long as $\xi h \ll 1$, and the result also turns out to be the same as Eq. (2.2.10) derived from the Rayleigh-Lamb solution for the limiting case of small ω and ξ. For this limiting case, therefore, an identical result is obtained from the classical plate theory, with or without rotatory inertia, as well as from the exact elasticity theory.

As ω and ξ increase, results of the classical plate equations begin to deviate from those of the elasticity theory. The accuracy of the classical plate equations can be improved by making use of the factor κ_r, as suggested by Mindlin. We recall that, for large ξ, the Rayleigh-Lamb solution for an infinite plate becomes the governing equation (2.2.11) for the phase velocity of Rayleigh's surface waves. The phase velocity was found to be, for instance,

$$c = \frac{\omega}{\xi} = 0.932\sqrt{\frac{E}{2\rho(1+\nu)}} \quad \text{for} \quad \nu = \frac{1}{3}.$$

As $\xi h \to \infty$, the corresponding phase velocity given by Eq. (3.5.2) is

$$c = \frac{\omega}{\xi} = \sqrt{\frac{E}{\kappa_r \rho(1-\nu^2)}}.$$

If κ_r is used as a correction factor, determined by matching the above two values of c, we find

$$\kappa_r = 3.46 \quad \text{for} \quad \nu = \frac{1}{3}.$$

For $\nu = 1/3$, results for the lowest branch of the frequency spectrum of flexural vibration of an infinite plate are shown in Figure 3.5.1, in which the ratio ω/ω_s is plotted versus ξh. The results include those given by the classical theory in Eqs. (3.5.2) and (3.5.3), as well as the exact Rayleigh-Lamb solution in Eq. (2.2.8). The highest curve in the figure is given by Eq. (3.5.3), does not include the effect of rotatory inertia, and is good for a frequency range of $\omega/\omega_s \leq 0.1$. By taking $\kappa_r = 1$ in Eq. (3.5.2), the rotatory inertia effect is taken into account, but no matching of the phase velocity is made. The result is given by the next lower curve and is good for a greater frequency range of $\omega/\omega_s \leq 0.2$. Finally, with $\kappa_r = 3.46$, matching of the phase velocity is made for $\nu = 1/3$, and the result given by Eq. (3.5.2) is in excellent agreement with the exact result for a much greater frequency range of $\omega/\omega_s \leq 1$, as is shown by the lowest curve.

To gain a feeling of the frequency range that is being discussed here, consider the example of a plate with a thickness of 1 cm. When the plate is made of steel, the value of ω_s is on the order of 10^6 radians per second, which is regarded as a high frequency in the vibration analysis of ordinary engineering structures.

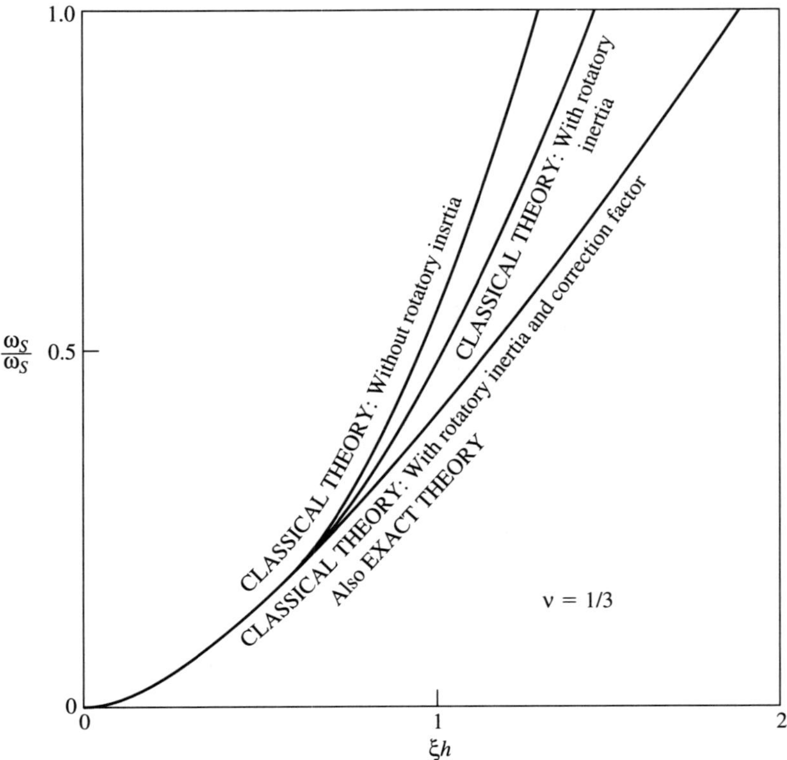

Fig. 3.5.1. Lowest branch of frequency spectrum of flexural vibration of an infinite plate.

3.5.2   Flexural Vibration Based on Refined Equations

A similar analysis can be carried out on the basis of the refined flexural equations (3.2.10). For plane–strain modes of free vibration in an infinite plate with dependence on x, these reduce to

$$D\frac{\partial^2 \psi}{\partial x^2} - 2\kappa G h \left(\psi + \frac{\partial w}{\partial x}\right) - \frac{2\rho h^3}{3}\ddot{\psi} = 0$$

$$2\kappa G h \left(\frac{\partial \psi}{\partial x} + \frac{\partial^2 w}{\partial x^2}\right) - 2\rho h \ddot{w} = 0. \tag{3.5.4}$$

By taking

$$w = W \sin \xi x \, e^{i\omega t}, \quad \psi = \Psi \cos \xi x \, e^{i\omega t}, \tag{3.5.5}$$

the following frequency equation is obtained:

$$\frac{1}{\kappa}\left(\frac{\pi}{2}\right)^4 \left(\frac{\omega}{\omega_s}\right)^4 - \left\{3 - \left[1 + \frac{2}{\kappa(1-\nu)}\right](\xi h)^2\right\}\left(\frac{\pi}{2}\right)^2 \left(\frac{\omega}{\omega_s}\right)^2$$

$$+ \frac{2}{1-\nu}(\xi h)^4 = 0, \tag{3.5.6}$$

which contains the dimensionless quantities ω/ω_s and ξh. For a given value of ξh, this now yields two frequencies instead of just one, as predicted by the classical plate equation. For small ξh, the lower frequency corresponds to a mode that is essentially flexural, and the higher frequency is for a predominantly thickness-shear mode. For $\xi h \ll 1$, Eqs. (3.5.4) yield, with $\kappa = \pi^2/12$,

$$\omega = \xi^2 h \sqrt{\frac{E}{3\rho(1 - \nu^2)}}, \qquad \omega = \omega_s.$$

The first of these frequencies is the same as Eq. (3.5.3) given by the classical equation. The second is the same as the lowest simple thickness-shear frequency given by the exact solution in elasticity; the value of $\kappa = \pi^2/12$ has indeed led to a perfect matching between the plate and elasticity solutions.

Numerical results for $\nu = 1/3$ are shown in Figure 3.5.2. In addition to the flexural branch, Eq. (3.5.6) now also yields the next higher branch, which is predominantly thickness-shear. Within the ranges of frequency and wavelength covered by the figure, the result for the flexural branch given by the refined theory is only

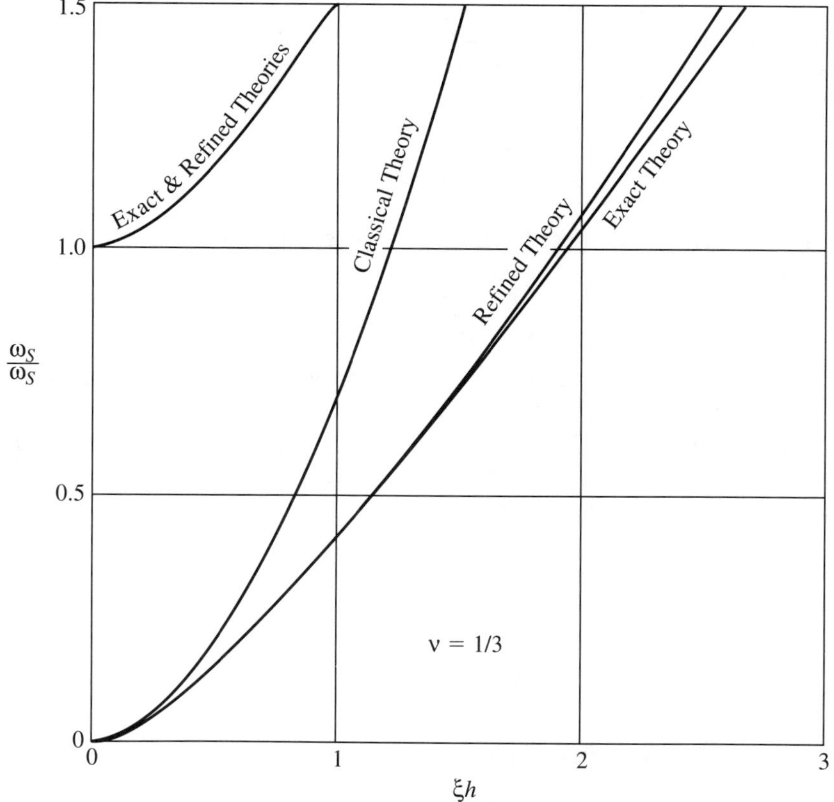

Fig. 3.5.2. Lowest two branches (flexural and thickness-shear) of frequency spectrum of flexural vibration of an infinite plate.

slightly higher than that given by the exact theory. In fact, the results are essentially identical up to the values of $\omega/\omega_s = 0.5$ and $\xi h = 1$. As was pointed out by Mindlin, these observations also agree with his experimental results. To serve as a comparison, the classical result given by Eq. (3.5.3) also has been included in the same figure.

To assess the effects of transverse shear and rotatory inertia, we rewrite the frequency equation (3.5.6) in the form

$$\frac{2\rho h^3/3}{2\kappa Gh}\omega^4 - \left[1 + \xi^2 \frac{D}{2\kappa Gh} + \xi^2 \frac{2\rho h^3/3}{2\rho h}\right]\omega^2$$
$$+ \frac{D}{2\rho h}\xi^4 = 0, \tag{3.5.7}$$

where the various quantities are grouped in the same manner as they have appeared in Eqs. (3.5.6). In the present form, the transverse shear effect may be suppressed readily by setting κ or $2\kappa Gh$ equal to infinity, in which case Eq. (3.5.7) reduces to the classical result in Eq. (3.5.2) that still includes the effect of rotatory inertia. Rotatory inertia may further be eliminated by setting $2\rho h^3/3$ equal to 0, in which case Eq. (3.5.2) becomes Eq. (3.5.3), as was mentioned earlier. Indeed, comparing Eq. (3.5.7) with (3.5.2) and (3.5.3), we conclude that, as far as the lower flexural mode is concerned, the transverse shear effect becomes negligible if the following conditions are fulfilled:

$$\frac{2\rho h^3/3}{2\kappa Gh}\omega^2 \ll 1, \quad \frac{D}{2\kappa Gh}\xi^2 \ll 1, \tag{3.5.8}$$

and the rotatory inertia effect becomes negligible if the following condition is fulfilled:

$$\frac{2\rho h^3/3}{2\rho h}\xi^2 \ll 1. \tag{3.5.9}$$

However, if the higher mode is to be retained in the analysis, rotatory inertia cannot be ignored for mathematical reasons.

The conditions in Eqs. (3.5.8) and (3.5.9) also can be expressed in terms of the dimensionless quantities ω/ω_s and ξh that have appeared in Eq. (3.5.6), in the form

$$\left(\frac{\omega}{\omega_s}\right)^2 \ll 1 \quad \text{and} \quad (\xi h)^2 \ll 1. \tag{3.5.10}$$

For instance, these may be interpreted as requiring

$$\frac{\omega}{\omega_s} < 0.1 \quad \text{and} \quad \xi h < 0.1. \tag{3.5.11}$$

For a steel plate with a thickness of 1 cm considered earlier, for which ω_s is of the order of 10^6 radians per second, the first of these conditions requires that the frequency be lower than 10^5 radians per second. Since $\xi h = (\pi/2)(2h/\ell)$ is of the order of the ratio between the plate thickness and the half-wavelength, the second of these conditions requires that this ratio be less than 0.1. These are the conditions under which the effects of transverse shear and rotatory inertia can be neglected and the simpler classical plate equations can be used instead of the refined equations.

The refined equations for plate flexure also yield the lowest horizontal shear mode of an infinite plate. By taking this mode in the form

$$\psi = \Psi \cos \eta y\, e^{i\omega t}, \quad \phi = w = 0,$$

Eqs. (3.2.10) yield the frequency equation

$$\left(\frac{\omega}{\omega_s}\right)^2 = 1 + \left(\frac{2\eta h}{\pi}\right)^2.$$

This agrees with the exact result based on elasticity for $q = 1$. The cutoff frequency of the horizontal shear branch is the same as that of the thickness-shear branch. For equal ξ and η, the latter is always higher.

3.5.3 Extensional Vibration

Equations (3.3.9) are the classical equations for plate extension. To determine the useful range, we ignore the external force terms as before and let

$$u = U \sin \xi x\, e^{i\omega t}, \quad v = 0. \tag{3.5.12}$$

The following frequency is obtained:

$$\omega = \xi \sqrt{\frac{E}{\rho(1 - v^2)}} \tag{3.5.13}$$

which is the same as Eq. (2.2.9) derived from the exact Rayleigh–Lamb solution for small ω and ξ.

When the refined equations (3.4.8) for extension are adopted, we let

$$u = U \sin \xi x\, e^{i\omega t}, \quad v = 0, \quad \beta = B \cos \xi x\, e^{i\omega t}. \tag{3.5.14}$$

The frequency equation is then

$$\left[\left(\frac{\pi}{2}\right)^2 \left(\frac{\omega}{\omega_s}\right)^2 \left(\frac{1}{c}\right)^2 - (\xi h)^2\right]\left\{\frac{1}{3c^2}\left[\frac{\pi^2}{4}\left(\frac{\omega}{\omega_s}\right)^2 - (\xi h)^2\right] - \kappa'\right\}$$

$$- \kappa'\left(1 - \frac{2}{c^2}\right)(\xi h)^2 = 0, \tag{3.5.15}$$

where c is the same as introduced in Eq. (2.3.11). For a given ξh, this equation gives two frequencies that may be compared with those given by the exact solution for the two lowest symmetric modes. For small ξh, the mode with a lower frequency is essentially face-extensional, and that with a higher frequency is essentially a thickness-stretch mode.

For $\xi h \ll 1$, the two frequencies given by Eq. (3.5.15) reduce to, with $\kappa' = \pi^2/12$,

$$\omega = \xi \sqrt{\frac{E}{\rho(1 - \nu^2)}}, \qquad \omega = c\omega_s. \tag{3.5.16}$$

The lower of these frequencies is the same as that given by the classical equations in Eq. (3.5.13) and thus also the same as that given by the exact solution in Eq. (2.2.9). The higher frequency in Eq. (3.5.16) cannot be derived from the classical equations. It is the same cutoff frequency of the lowest symmetric thickness-stretch mode given by the exact solution, but only when $\nu < 1/3$. According to the exact solution, when $\nu = 1/3$, the frequency of the lowest symmetric thickness- stretch mode becomes equal to that of the lowest symmetric thickness-shear mode. When $\nu > 1/3$, the latter turns out to be lower than the former. Since symmetric thickness-shear motion is not included in the present refined theory of extension, it is not valid for $\nu > 1/3$, as was pointed out in Section 3.4.

Numerical results for the frequencies of the first two symmetric modes are shown in Figure 3.5.3 for $\nu = 1/3$. Although the classical plate solution produces only the lowest mode, the result is essentially the same as that given by the refined plate solution or the exact elasticity solution for values of the coordinates

$$\xi h < 1 \quad \text{and} \quad \frac{\omega}{\omega_s} < 1.$$

Within these ranges, therefore, the classical plate equations for extension are practically as good as the refined plate equations and the exact elasticity equations. To develop still better plate equations for extension for higher frequencies, Mindlin and Medick (1959) had to include the lowest symmetric thickness-shear mode, as was mentioned earlier.

3.5.4   Vibrations with Phase Reversals in Both x- and y-Directions

The plane–strain vibration in an infinite plate may be extended readily to vibrations with phase reversals in both the x- and y-directions. Instead of the one-dimensional waveforms in Eqs. (3.5.1), (3.5.5), (3.5.12), and (3.5.14), we now adopt a more general two-dimensional form. A typical example is given by the plate deflection

$$w = W \sin \xi_x x \sin \xi_y y \, e^{i\omega t} \tag{3.5.17}$$

with

$$\xi_x = \frac{2\pi}{a}, \quad \xi_y = \frac{2\pi}{b},$$

where a and b are the wavelengths in the x- and y-directions, respectively. By substituting plate displacements such as given by Eq. (3.5.17) into the displacement equations of motion of the plate, a frequency equation may be derived in the usual

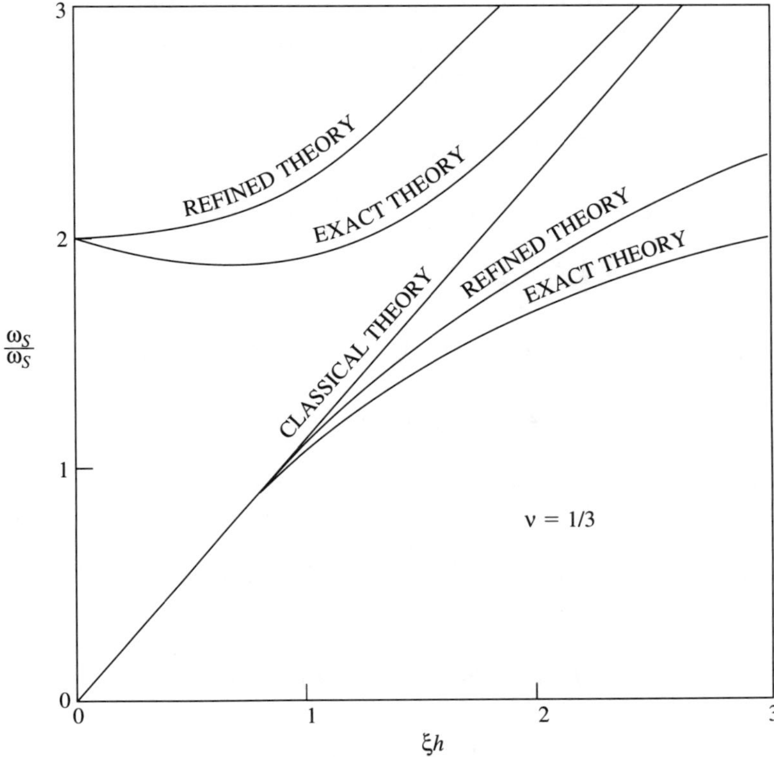

Fig. 3.5.3. Lowest two branches of frequency spectrum of extensional vibration of an infinite plate.

manner. For flexural vibrations in an infinite plate based on the refined equations, the frequency equation still gives the flexural and thickness-shear branches, but the horizontal-shear branch becomes a thickness-twist branch. The results are similar to those for flexural vibrations in a simply supported rectangular plate (Mindlin *et al.* 1956).

3.6   General Equations of an Anisotropic Plate

A generalized Hamilton's principle and the associated variational equation of motion in nonlinear elasticity theory have been presented in Section 1.6. A linearized version of these is deduced in this section. The linear variational equation of motion is then employed in the derivation of a system of general equations of an anisotropic plate (Yu 1965) by expanding the displacements and strains in infinite power series in the manner of Cauchy and Mindlin (Mindlin 1961). The final results consist of a complete system of plate equations of all orders which are essentially similar to those given by Mindlin.

3.6.1 Generalized Hamilton's Principle in Linear Elasticity

For a linear, anisotropic elastic solid, we write the generalized Hamilton's principle in Eqs. (1.6.1) through (1.6.5) in the form

$$\delta \int_{t_0}^{t_1} L \, dt = \delta \int_{t_0}^{t_1} (T - U + W) \, dt = 0, \tag{3.6.1}$$

where

$$T = \int \tfrac{1}{2} \rho \dot{u}_i \dot{u}_i \, dV$$

$$U = \int \left[\tfrac{1}{2} \sigma_{ij}(u_{i,j} + u_{j,i}) - \sigma_{ij} e_{ij} + \tfrac{1}{2} c_{ijkl} e_{ij} e_{kl} \right] dV \tag{3.6.2}$$

$$W = \int f_i u_i \, dV + \int \bar{p}_i u_i \, dS_p + \int p_i (u_i - \bar{u}_i) \, dS_u.$$

In these equations, c_{ijkl} is the elastic stiffness tensor, an overbar denotes a prescribed quantity, and other notations are similar to those in Eqs. (1.6.2) through (1.6.5).

When variations of displacements, strains, and stresses are taken independently and simultaneously, Eq. (3.6.1) yields

$$\int_{t_0}^{t_1} dt \int (\sigma_{ij,i} + f_j - \rho \ddot{u}_j) \, \delta u_j \, dV$$

$$- \int_{t_0}^{t_1} dt \int (\sigma_{ij} \nu_i - \bar{p}_j) \, \delta u_j \, dS_p$$

$$- \int_{t_0}^{t_1} dt \int (\sigma_{ij} - c_{ijkl} e_{kl}) \, \delta e_{ij} \, dV \tag{3.6.3}$$

$$- \int_{t_0}^{t_1} dt \int \left[e_{ij} - \tfrac{1}{2}(u_{i,j} + u_{j,i}) \right] \delta \sigma_{ij} \, dV$$

$$- \int_{t_0}^{t_1} dt \int (u_i - \bar{u}_i) \, \delta p_i \, dS_u = 0,$$

which is the linearized version of the generalized variational equation of motion. The Euler equations written from Eq. (3.6.3) give successively the stress equations of motion, traction boundary conditions, stress–strain relations, strain–displacement relations, and displacement boundary conditions, which constitute the complete system of equations of linear anisotropic elasticity.

3.6.2 Equations of an Anisotropic Plate

For the plate in Figure 3.1.1, we follow Mindlin by taking

$$u_i = \sum_{n=0}^{\infty} z^n u_i^{(n)}, \quad e_{ij} = \sum_{n=0}^{\infty} z^n e_{ij}^{(n)}, \quad \sigma_{ij}^{(n)} = \int_{-h}^{h} \sigma_{ij} z^n \, dz, \tag{3.6.4}$$

where $u_i^{(n)}$, $e_{ij}^{(n)}$, and $\sigma_{ij}^{(n)}$ are independent of the thickness coordinate z and represent the nth-order components of the plate displacement, strain, and stress, respectively.

We substitute u_i and e_{ij} from Eqs. (3.6.4) into (3.6.2), carry out the integration with respect to z over the thickness of the plate, and introduce $\sigma_{ij}^{(n)}$ again according to Eqs. (3.6.4). The results are

$$T = \int \frac{1}{2}\rho \sum_{m=0}^{\infty} \sum_{n=0}^{\infty} H_{mn}\, \ddot{u}_i^{(m)} \ddot{u}_i^{(n)}\, dA$$

$$U = \int \left\{ \frac{1}{2} \sum_{n=0}^{\infty} \sigma_{ij}^{(n)} \left[\left(u_i^{(n)}\right)_{,j} + \left(u_j^{(n)}\right)_{,i} + (n+1)\left(\delta_{xj}u_i^{(n+1)} + \delta_{ix}u_j^{(n+1)}\right)\right] \right.$$

$$\left. - \sum_{n=0}^{\infty} \sigma_{ij}^{(n)} e_{ij}^{(n)} + \frac{1}{2}c_{ijkl} \sum_{m=0}^{\infty}\sum_{n=0}^{\infty} H_{mn} e_{ij}^{(n)} e_{kl}^{(m)} \right\} dA \qquad (3.6.5)$$

$$W = \int \sum_{n=0}^{\infty} f_i^{(n)} u_i^{(n)}\, dA + \int \sum_{n=0}^{\infty} \bar{P}_i^{(n)} u_i^{(n)}\, dA_p + \int \sum_{n=0}^{\infty} P_i^{(n)} (u_i^{(n)} - \bar{u}_i^{(n)})\, dA_u$$

$$+ \int \sum_{n=0}^{\infty} \bar{p}_i^{(n)} u_i^{(n)}\, dC_p + \int \sum_{n=0}^{\infty} p_i^{(n)} (u_i^{(n)} - \bar{u}_i^{(n)})\, dC_u,$$

where

$$H_{mn} = \frac{2h^{m+n+1}}{m+n+1} \qquad \text{for even } (m+n)$$

$$= 0 \qquad \text{for odd } (m+n)$$

$$f_i^{(n)} = \int_{-h}^{h} f_i z^n\, dz$$

$$p_i^{(n)} = \int_{-h}^{h} p_i z^n\, dz$$

$$P_i^{(n)} = [p_i z^n]_{z=h} + [p_i z^n]_{z=-h}.$$

In addition, δ_{xj} and δ_{ix} are Kronecker deltas, A_p and A_u are those parts of the plan area A of the plate, and C_p and C_u are those parts of the edge of the plate on which traction and displacement, respectively, are prescribed. The generalized Hamilton's principle in Eq. (3.6.1) becomes that of a plate when Eqs. (3.6.5) take the place of Eqs. (3.6.2).

The generalized variational equation of motion for the plate now may be derived from Eqs. (3.6.1) with (3.6.5) or, more directly, by carrying out integration with respect to z in Eq. (3.6.3) together with the use of Eqs. (3.6.4). In either case, the result is

$$\int_{t_0}^{t_1} dt \int \sum_{n=0}^{\infty} \left[\sigma_{ij,i}^{(n)} - n\sigma_{xj}^{(n-1)} + \bar{p}_j^{(n)} + f_j^{(n)} - \rho \sum_{m=0}^{\infty} H_{mn}\ddot{u}_j^{(m)} \right] \delta u_i^{(n)}\, dA$$

$$+ \int_{t_0}^{t_1} dt \int \sum_{n=0}^{\infty} \left[\sigma_{ij}^{(n)} - c_{ijkl} \sum_{m=0}^{\infty} H_{mn} e_{kl}^{(m)} \right] \delta e_{ij}^{(n)}\, dA$$

$$+ \int_{t_0}^{t_1} dt \int \sum_{n=0}^{\infty} \left\{ e_{ij}^{(n)} - \frac{1}{2} \left[(u_i^{(n)})_{,j} + (u_j^{(n)})_{,i} \right. \right.$$

$$\left. \left. + (n+1)(\delta_{xj} u_i^{(n+1)} + \delta_{ix} u_j^{(n+1)}) \right] \right\} \delta\sigma_{ij}^{(n)} \, dA$$

$$+ \int_{t_0}^{t_1} dt \int \sum_{n=0}^{\infty} (u_i^{(n)} - \bar{u}_i^{(n)}) \delta P_i^{(n)} \, dA_u$$

$$- \int_{t_0}^{t_1} dt \int \sum_{n=0}^{\infty} [(\sigma_{nn}^{(n)} - \bar{p}_n^{(n)}) \delta u_n^{(n)} + (\sigma_{ns}^{(n)} - \bar{p}_s^{(n)}) \delta u_s^{(n)}$$

$$+ (\sigma_{nz}^{(n)} - \bar{p}_z^{(n)}) \delta u_z^{(n)}] \, dC_p$$

$$+ \int_{t_0}^{t_1} dt \int \sum_{n=0}^{\infty} [(u_n^{(n)} - \bar{u}_n^{(n)}) \delta p_n^{(n)} + (u_s^{(n)} - \bar{u}_s^{(n)}) \delta p_s^{(n)}$$

$$+ (u_z^{(n)} - \bar{u}_z^{(n)}) \delta p_z^{(n)}] \, dC_u = 0, \tag{3.6.6}$$

where the subscripts n and s denote directions normal and tangential to the plate contour.

The Euler equations written from Eq. (3.6.6) constitute the complete system of equations of an anisotropic plate. In addition to the stress equations of motion and plate stress–strain–displacement relations, traction and displacement boundary conditions around the plate contour are provided. Tractions at the top and bottom boundary planes of the plate as well as body forces also can be prescribed. When variations are restricted to those of displacements, the ordinary variational equation of motion of the plate is deduced from Eq. (3.6.6) as a special case.

Through truncation of the infinite series, the equations of an anisotropic plate written from Eqs. (3.6.6) are reducible to Mindlin's equations of a homogeneous plate (Mindlin 1951a) and a crystal plate (Mindlin 1951b) as special cases.

References

Kane, T.R. and R.D. Mindlin (1956) High-Frequency Extensional Vibrations of Plates. *Journal of Applied Mechanics*, Vol. 23, pp. 277–283.

Mindlin, R.D. (1951a) Influence of Rotatory Inertia and Shear on Flexural Motions of Isotropic, Elastic Plates. *Journal of Applied Mechanics*, Vol. 18, pp. 31–38.

Mindlin, R.D. (1951b) Thickness-Shear and Flexural Vibrations of Crystal Plates. *Journal of Applied Physics*, Vol. 22, pp. 316-323.

Mindlin, R.D. (1961) High Frequency Vibrations of Crystal Plates. *Quarterly of Applied Mathematics*, Vol. 19, pp. 51–61.

Mindlin, R.D., A. Schacknow, and H. Deresiewicz (1956) Flexural Vibrations of Rectangular Plates. *Journal of Applied Mechanics*, Vol. 23, pp. 430–436.

Mindlin, R.D. and M.A. Medick (1959) Extensional Vibrations of Elastic Plates. *Journal of Applied Mechanics*, Vol. 26, pp. 561–569.

Reissner, E. (1945) The Effect of Transverse Shear Deformation on the Bending of Elastic Plates. *Journal of Applied Mechanics*, Vol. 67, pp. A-69–A-77.

Timoshenko, S. (1921) On the Correction for Shear of the Differential Equation for Transverse Vibrations of Prismatic Bars. *Philosophical Magazine*, Vol. 41, pp. 744–746.

Timoshenko, S. and S. Woinowsky-Krieger (1959). *Theory of Plates and Shells*, 2nd Ed. McGraw-Hill, New York.

Yu, Y.Y. (1965) On Linear Equations of Anisotropic Elastic Plates. *Quarterly of Applied Mathematics*, Vol. 22, pp. 357–360.

Yu, Y.Y. (1992) Equations for Large Deflections of Elastic and Piezoelectric Plates and Shallow Shells, Including Sandwiches and Laminated Composites, with Applications to Vibrations, Chaos, and Acoustic Radiation. Presented at the XIIIth International Congress of Theoretical and Applied Mechanics, Haifa, Israel.

Yu, Y.Y. (1995) On the Ordinary, Generalized, and Pseudo-Variational Equations of Motion in Nonlinear Elasticity, Piezoelectricity, and Classical Plate Theories. *Journal of Applied Mechanics*, Vol. 62, pp. 471–478.

4

Linear Modeling of Sandwich Plates

The sandwich plate that was analyzed in Chapter 2 on the basis of linear elasticity theory is treated again in this chapter, now on the basis of equations of sandwich plates to be derived. We recall that the sandwich plate consists of three homogeneous, elastic, and isotropic layers that are symmetrically constructed with respect to the middle plane of the sandwich. The two face layers are thus identical, although the core of the sandwich is made of a different material and can have a different thickness. Other than being symmetrically constructed, the sandwich is therefore an arbitrary three-layered plate, including as an important special case the ordinary sandwich plate that has relatively thin but rigid and heavy face layers. Being symmetric, flexural and extensional linear vibrations of the sandwich plate are uncoupled from each other.

Mindlin's approach (1951) to the linear vibration analysis of homogeneous plates was extended by Yu (1959a,b, 1960a,b) to develop a general system of equations for sandwich plates. According to Habip (1965), the latter author was among the first to investigate the vibrations of sandwich plates. More recently, Yu's work was credited by Toledano and Murakami (1987) with being the first to derive an approximate theory of layered plates and to adopt piecewise linear continuous displacements in such a theory. An important simplified version (Yu 1960c) was developed for ordinary sandwich plates for which the governing equations take a form analogous to Mindlin's equations for homogeneous plates.

In addition to Habip's review, early reviews of vibrations of layered plates and shells were also given by Knoell and Robinson (1975), Bert (1975), and others. A recent survey of ours has appeared both in English (Yu 1989) and in Russian (Yu 1992), and covers linear and nonlinear vibrations of not only sandwiches but also laminated composites. Since our first effort was made (Yu 1959a,b, 1960a,b), many publications on the vibrations of sandwich structures have appeared. Among the original contributions, the importance of transverse shear effect in sandwich has been shown in our early results (Yu 1960a) and further substantiated by Nicholas and Heller (1967). Our early work on a symmetric sandwich with membrane facings (Yu 1960b) was extended to an unsymmetric sandwich by Rao and Nakra (1974). Our other early works covered forced vibrations of sandwich beams (Yu 1960b), damping in sandwiches (Yu 1962a), and extensional vibrations of sandwiches (Yu 1962b).

In this chapter, we derive linear equations of sandwich plates, again from the variational equation of motion in linear elasticity. In Section 4.1, we introduce a general system of equations for flexure that include transverse shear effects in all layers in a three-layered plate. A system of simplified equations for flexure is next presented in Section 4.2 by considering the face layers as membranes in which the shear effect is no longer considered. In Section 4.3, classical equations of flexure are derived by suppressing the shear effect in all layers. Based on these three systems of equations, flexural vibrations of an infinite sandwich plate are then investigated in Section 4.4. In the last section in this chapter, equations for extension of a sandwich plate are similarly derived and applied to the analysis of extensional vibrations of an infinite sandwich plate.

4.1   Refined Equations for Flexure of a Sandwich Plate Including Transverse Shear Effects in All Layers

As was shown before in Figure 2.6.1, we consider a symmetrically constructed isotropic elastic sandwich plate. The two face layers are identical, but no restrictions are imposed upon the ratios between the thicknesses, material densities, and elastic constants of the core and face layers (Yu, 1959a, 1960a, 1962a). Perfect bonds are assumed to exist at the interfaces between the face and core layers, which requires the continuity of both displacements and stresses at the interfaces.

As before, the middle plane of the sandwich plate is chosen to be the xy-plane. In the z- direction, the thicknesses of the lower face, core, and upper face layers extend from $-h$ to $-h_1$, $-h_1$ to h_1, and h_1 to h, respectively. The thickness of the core is thus $2h_1$, and that of each of the two face layers is $h_2 = h - h_1$. The layers are identified through the use of the subscript i, with $i = 1, 2, 3$ denoting the core, lower face, and upper face layers, respectively. When only the values 1 and 2 are used for the subscript, they refer to the core layer and face layers, respectively.

4.1.1   Variational Equation of Motion

To include transverse shear effects in all three layers of the sandwich, we assume the three- dimensional displacements in the form

$$u_{x1} = z\,\psi_1, \qquad u_{x2}, u_{x3} = \mp h_1(\psi_1 - \psi_2) + z\,\psi_2$$

$$u_{y1} = z\,\phi_1, \qquad u_{y2}, u_{y3} = \mp h_1(\phi_1 - \phi_2) + z\phi_2 \qquad (4.1.1)$$
$$u_{z1} = u_{z2} = u_{z3} = w.$$

Following the procedure used in the preceding chapter, we substitute Eqs. (4.1.1) into the linear variational equation (1.4.9) and carry out integration with respect to z over the entire thickness of the sandwich, which now includes the thicknesses of all three layers. The result is

$$
\int_{t_0}^{t_1} dt \int \int_A \left\{ \left[\frac{\partial}{\partial x}\{M_{x1} + h_1(N_{x3} - N_{x2})\} \right.\right.
$$
$$
+ \frac{\partial}{\partial y}\{M_{yx1} + h_1(N_{yx3} - N_{yx2})\}
$$
$$
- Q_{x1} + h_1(p_x^+ - p_x^-) + f_{x1}^{(1)} + h_1(f_{x3}^{(0)} - f_{x2}^{(0)})
$$
$$
\left. - \left(\frac{2}{3}\rho_1 h_1^3 + 2\rho_2 h_2 h_1^2\right)\ddot{\psi}_1 - \rho_2 h_1 h_2^2 \ddot{\psi}_2 \right]\delta\psi_1
$$
$$
+ \left[\frac{\partial}{\partial x}\{M_{x2} + M_{x3} - h_1(N_{x3} - N_{x2})\} \right.
$$
$$
+ \frac{\partial}{\partial y}\{M_{yx2} + M_{yx3} - h_1(N_{yx3} - N_{yx2})\}
$$
$$
- Q_{x2} - Q_{x3} + h_2(p_x^+ - p_x^-) + f_{x2}^{(1)} + f_{x3}^{(1)} - h_1\left(f_{x3}^{(0)} - f_{x2}^{(0)}\right)
$$
$$
\left. - \rho_2 h_1 h_2^2 \ddot{\psi}_1 - \frac{2}{3}\rho_2 h_2^3 \ddot{\psi}_2 \right]\delta\psi_2
$$
$$
+ \left[\frac{\partial}{\partial x}\{M_{xy1} + h_1(N_{xy3} - N_{xy2})\} + \frac{\partial}{\partial y}\{M_{y1} + h_1(N_{y3} - N_{y2})\} \right.
$$
$$
- Q_{y1} + h_1(p_y^+ - p_y^-) + f_{y1}^{(1)} + h_1(f_{y3}^{(0)} - f_{y2}^{(0)})
$$
$$
\left. - \left(\frac{2}{3}\rho_1 h_1^3 + 2\rho_2 h_2 h_1^2\right)\ddot{\phi}_1 - \rho_2 h_1 h_2^2 \ddot{\phi}_2 \right]\delta\phi_1
$$
$$
+ \left[\frac{\partial}{\partial x}\{M_{xy2} + M_{xy3} - h_1(N_{xy3} - N_{xy2})\} \right.
$$
$$
+ \frac{\partial}{\partial y}\{M_{y2} + M_{y3} - h_1(N_{y3} - N_{y2})\}
$$
$$
- Q_{y2} - Q_{y3} + h_2(p_y^+ - p_y^-) + f_{y2}^{(1)} + f_{y3}^{(1)} - h_1(f_{y3}^{(0)} - f_{y2}^{(0)})
$$
$$
\left. - \rho_2 h_1 h_2^2 \ddot{\phi}_1 - \frac{2}{3}\rho_2 h_2^3 \ddot{\phi}_2 \right]\delta\phi_2
$$
$$
+ \left[\frac{\partial}{\partial x}(Q_{x1} + Q_{x2} + Q_{x3}) + \frac{\partial}{\partial y}(Q_{y1} + Q_{y2} + Q_{y3}) + p_z^+ + p_z^- \right.
$$

$$\left. +f_{z1}^{(0)} + f_{z2}^{(0)} + f_{z3}^{(0)} - 2(\rho_1 h_1 + \rho_2 h_2)\ddot{w}\right] \delta w\right\} dx\,dy$$

$$-\int_{t_0}^{t_1} dt \oint_{C_p} \{[M_{n1} + h_1(N_{n3} - N_{n2}) - p_{n1}^{(1)} - h_1(p_{n3}^{(0)} - p_{n2}^{(0)})]\delta\psi_{n1}$$

$$+ [M_{n2} + M_{n3} - h_1(N_{n3} - N_{n2}) - p_{n2}^{(1)} - p_{n3}^{(1)}$$

$$+ h_1(p_{n3}^{(0)} - p_{n2}^{(0)})]\delta\psi_{n2}$$

$$+ [M_{ns1} + h_1(N_{ns3} - N_{ns2}) - p_{s1}^{(1)} - h_1(p_{s3}^{(0)} - p_{s2}^{(0)})]\delta\psi_{s1}$$

$$+ [M_{ns2} + M_{ns3} - h_1(N_{ns3} - N_{ns2}) - p_{s2}^{(1)} - p_{s3}^{(1)}$$

$$+ h_1(p_{s3}^{(0)} - p_{s2}^{(0)})]\delta\psi_{s2}$$

$$+ [Q_{n1} + Q_{n2} + Q_{n3} - p_{z1}^{(0)} - p_{z2}^{(0)} - p_{z3}^{(0)}]\delta w\} \, ds = 0. \tag{4.1.2}$$

In Eq. (4.1.2), the plate stresses are defined by

$$[M_{xi}, M_{yi}, M_{xyi}] = \int_{a_i}^{b_i} [\sigma_{xxi}, \sigma_{yyi}, \sigma_{xyi}]\, z\, dz$$

$$[N_{xi}, N_{yi}, N_{xyi}] = \int_{a_i}^{b_i} [\sigma_{xxi}, \sigma_{yyi}, \sigma_{xyi}]\, dz \tag{4.1.3}$$

$$[Q_{xi}, Q_{yi}] = \int_{a_i}^{b_i} [\sigma_{zxi}, \sigma_{zyi}]\, dz,$$

which are the usual bending and twisting moments, membrane forces, and transverse shearing forces, respectively, and a_i and b_i are values of z for the lower and upper boundary planes of each layer. Thus,

$$a_1 = -h_1, \quad b_1 = +h_1; \quad a_2 = -h, \quad b_2 = -h_1;$$
$$a_3 = +h_1, \quad b_3 = +h.$$

Similarly, the plate body forces and plate edge tractions are defined by, respectively,

$$\left[f_{xi}^{(0)}, f_{yi}^{(0)}, f_{xi}^{(1)}, f_{yi}^{(1)}\right] = \int_{a_i}^{b_i} \left[f_{xi}, f_{yi}, f_{xi}z, f_{yi}z\right]\, dz$$

$$\left[p_{ni}^{(0)}, p_{si}^{(0)}, p_{zi}^{(0)}, p_{ni}^{(1)}, p_{si}^{(1)}\right] = \int_{a_i}^{b_i} \left[p_{ni}, p_{si}, p_{zi}, p_{ni}z, p_{si}z\right]\, dz.$$

Finally, n and s refer to directions normal and tangential to the plate contour; p_x^+ and p_x^- are surface tractions in the x-direction at the top and bottom boundary planes $z = \pm h$, and those in the y- and z-directions are similar.

Equation (4.1.2) is the two-dimensional variational equation of motion for the sandwich plate from which refined equations may be written in the same manner as before. Thus, the stress equations of motion are, from the integral over the area A,

$$\frac{\partial}{\partial x}\{M_{x1} + h_1(N_{x3} - N_{x2})\} + \frac{\partial}{\partial y}\{M_{yx1} + h_1(N_{yx3} - N_{yx2})\}$$

$$- Q_{x1} + h_1(p_x^+ - p_x^-) + f_{x1}^{(1)} + h_1(f_{x3}^{(0)} - f_{x2}^{(0)})$$

$$-\left(\frac{2}{3}\rho_1 h_1^3 + 2\rho_2 h_2 h_1^2\right)\ddot{\psi}_1 - \rho_2 h_1 h_2^2 \ddot{\psi}_2 = 0$$

$$\frac{\partial}{\partial x}\{M_{x2} + M_{x3} - h_1(N_{x3} - N_{x2})\} + \frac{\partial}{\partial y}\{M_{yx2} + M_{yx3} - h_1(N_{yx3} - N_{yx2})\}$$

$$-Q_{x2} - Q_{x3} + h_2(p_x^+ - p_x^-) + f_{x2}^{(1)} + f_{x3}^{(1)} - h_1(f_{x3}^{(0)} - f_{x2}^{(0)})$$

$$- \rho_2 h_1 h_2^2\,\ddot{\psi}_1 - \frac{2}{3}\rho_2 h_2^3\,\ddot{\psi}_2 = 0$$

$$\frac{\partial}{\partial x}\{M_{xy1} + h_1(N_{xy3} - N_{xy2})\} + \frac{\partial}{\partial y}\{M_{y1} + h_1(N_{y3} - N_{y2})\} \qquad (4.1.4)$$

$$-Q_{y1} + h_1(p_y^+ - p_y^-) + f_{y1}^{(1)} + h_1(f_{y3}^{(0)} - f_{y2}^{(0)})$$

$$- \left(\tfrac{2}{3}\rho_1 h_1^3 + 2\rho_2 h_2 h_1^2\right)\ddot{\phi}_1 - \rho_2 h_1 h_2^2 \ddot{\phi}_2 = 0$$

$$\frac{\partial}{\partial x}\{M_{xy2} + M_{xy3} - h_1(N_{xy3} - N_{xy2}\} + \frac{\partial}{\partial y}\{M_{y2} + M_{y3} - h_1(N_{y3} - N_{y2})\}$$

$$-Q_{y2} - Q_{y3} + h_2(p_y^+ - p_y^-) + f_{y2}^{(1)} + f_{y3}^{(1)} - h_1(f_{y3}^{(0)} - f_{y2}^{(0)})$$

$$-\rho_2 h_1 h_2^2\,\ddot{\phi}_1 - \frac{2}{3}\rho_2 h_2^3\,\ddot{\phi}_2 = 0$$

$$\frac{\partial}{\partial x}(Q_{x1} + Q_{x2} + Q_{x3}) + \frac{\partial}{\partial y}(Q_{y1} + Q_{y2} + Q_{y3})$$

$$+ p_z^+ + p_z^- + f_{z1}^{(0)} + f_{z2}^{(0)} + f_{z3}^{(0)} - 2(\rho_1 h_1 + \rho_2 h_2)\ddot{w} = 0$$

and the boundary conditions are, from the boundary integral,

$$M_{n1} + h_1(N_{n3} - N_{n2}) - p_{n1}^{(1)} - h_1(p_{n3}^{(0)} - p_{n2}^{(0)}) = 0$$
$$\text{or} \quad \psi_{n1} = \text{prescribed}$$

$$M_{n2} + M_{n3} - h_1(N_{n3} - N_{n2}) - p_{n2}^{(1)} - p_{n3}^{(1)} + h_1(p_{n3}^{(0)} - p_{n2}^{(0)}) = 0$$
$$\text{or} \quad \psi_{n2} = \text{prescribed}$$

$$M_{ns1} + h_1(N_{ns3} - N_{ns2}) - p_{s1}^{(1)} - h_1(p_{s3}^{(0)} - p_{s2}^{(0)}) = 0 \qquad (4.1.5)$$
$$\text{or} \quad \psi_{s1} = \text{prescribed}$$

$$M_{ns2} + M_{ns3} - h_1(N_{ns3} - N_{ns2}) - p_{s2}^{(1)} - p_{s3}^{(1)} + h_1(p_{s3}^{(0)} - p_{s2}^{(0)}) = 0$$
$$\text{or} \quad \psi_{s2} = \text{prescribed}$$

$$Q_{n1} + Q_{n2} + Q_{n3} - p_{z1}^{(0)} - p_{z2}^{(0)} - p_{z3}^{(0)} = 0$$
$$\text{or} \quad w = \text{prescribed}.$$

4.1.2 Plate Stress–Displacement Relations

To transform the stress equations of motion into displacement equations of motion, we need relations between the plate stresses and plate displacements in each of the three layers. While the core layer as well as the sandwich plate as a whole undergoes only flexure, each of the two face layers involves both extension and flexure. Since transverse shear effects are included in all layers, the bending and twisting moments and transverse shearing forces can be written in a similar manner as those appearing in the refined equations of flexure of a homogeneous plate in Section 3.2. On the other hand, as thickness-stretch is not considered in the sandwich, the membrane forces can be treated in the same way as in the classical equations of extension of a homogeneous plate in Section 3.3.

Based on the results in Sections 3.2 and 3.3, the plate stresses are introduced in a manner similar to those for a homogeneous plate as follows:

$$[M_{x(i)},\ M_{y(i)},\ M_{xy(i)}] = \int_{-\delta_i/2}^{\delta_i/2} [\sigma_{xxi},\ \sigma_{yyi},\ \sigma_{xyi}]\, z_i\, dz_i$$

$$[N_{x(i)},\ N_{y(i)},\ N_{xy(i)}] = \int_{-\delta_i/2}^{\delta_i/2} [\sigma_{xxi},\ \sigma_{yyi},\ \sigma_{xyi}]\, dz_i \qquad (i = 1, 2, 3)$$

$$[Q_{x(i)},\ Q_{y(i)}] = \int_{-\delta_i/2}^{\delta_i/2} [\sigma_{zxi},\ \sigma_{zyi}]\, dz_i,$$

$$(4.1.6)$$

in which a subscript i enclosed in parentheses indicates that integration is carried out with respect to z_i, instead of z as in Eqs. (4.1.3). Relations between z_i and z are given by

$$z_1 = z, \qquad z_2 = z - h_1 - \frac{h_2}{2}, \qquad z_3 = z + h_1 + \frac{h_2}{2}.$$

The integration then covers the thickness of each layer, written as

$$\delta_i = b_i - a_i.$$

The plate stresses in Eqs. (4.1.3) are readily related to those in Eqs. (4.1.6). In fact, except for M_{x2} and M_{x3}, integration with respect to z or z_i yields the same result. The results for M_{x2} and M_{x3} are

$$M_{x2} = M_{x(2)} - \left(h_1 + \frac{h_2}{2}\right) N_{x(2)}$$

$$M_{x3} = M_{x(3)} + \left(h_1 + \frac{h_2}{2}\right) N_{x(3)}.$$

$$(4.1.7)$$

By making use of the results in Eqs. (3.2.7), (3.2.9), and (3.3.8), the plate stress–displacement relations are found to be given by

$$M_{x(i)} = D_i \left(\frac{\partial \psi_i}{\partial x} + v_i \frac{\partial \phi_i}{\partial y}\right)$$

$$M_{y(i)} = D_i \left(\frac{\partial \phi_i}{\partial y} + v_i \frac{\partial \psi_i}{\partial x}\right) \qquad (i = 1, 2, 3) \qquad (4.1.8)$$

$$M_{xy(i)} = \tfrac{1}{2} D_i (1 - v_i) \left(\frac{\partial \psi_i}{\partial y} + \frac{\partial \phi_i}{\partial x}\right)$$

$$Q_{x(i)} = \kappa_i G_i \delta_i \left(\psi_i + \frac{\partial w}{\partial x} \right)$$

$$Q_{y(i)} = \kappa_i G_i \delta_i \left(\phi_i + \frac{\partial w}{\partial y} \right)$$

$$(i = 1, 2, 3) \qquad (4.1.9)$$

$$N_{x(3)} = -N_{x(2)} = C_2 \left[h_1 \left(\frac{\partial \psi_1}{\partial x} + v_2 \frac{\partial \phi_1}{\partial y} \right) + \frac{h_2}{2} \left(\frac{\partial \psi_2}{\partial x} + v_2 \frac{\partial \phi_2}{\partial y} \right) \right]$$

$$N_{y(3)} = -N_{y(2)} = C_2 \left[h_1 \left(\frac{\partial \phi_1}{\partial y} + v_2 \frac{\partial \psi_1}{\partial x} \right) + \frac{h_2}{2} \left(\frac{\partial \phi_2}{\partial y} + v_2 \frac{\partial \psi_2}{\partial x} \right) \right]$$

$$N_{xy(3)} = -N_{xy(2)} \qquad (4.1.10)$$

$$= \tfrac{1}{2} C_2 (1 - v_2) \left[h_1 \left(\frac{\partial \psi_1}{\partial y} + \frac{\partial \phi_1}{\partial x} \right) + \frac{h_2}{2} \left(\frac{\partial \psi_2}{\partial y} + \frac{\partial \phi_2}{\partial x} \right) \right]$$

where

$$D_i = \frac{E_i}{1 - v_i^2} \frac{\delta_i^3}{12}, \qquad C_i = \frac{E_i \delta_i}{1 - v_i^2}$$

$$\delta_1 = 2h_1, \qquad \delta_2 = \delta_3 = h_2,$$

and κ_i are shear factors for the sandwich plate, to be determined presently. The final plate stress–displacement relations are obtained through the use of Eqs. (4.1.7) through (4.1.10). By means of these final results, the stress equations (4.1.4) are transformed readily into the displacement equations of motion of the sandwich plate. To save space, however, the complete displacement equations will not be written out here.

4.1.3   Determination of Shear Factors

The shear factors introduced in Eqs. (4.1.9) are determined by examining the simple thickness- shear modes of vibration of an infinite sandwich plate. Based on the exact linear elasticity theory, the frequency equation has been found to be given by Eq. (2.7.10a). Based on the present sandwich plate theory, the frequency equation is found by ignoring the surface and body forces and by letting

$$\psi_i = \Psi_1 \, e^{i\omega t}, \qquad \psi_2 = \Psi_2 \, e^{i\omega t}, \qquad \phi_1 = \phi_2 = w = 0$$

in the stress equations (4.1.4) and the plate stress–displacement relations in Eqs. (4.1.8) through (4.1.10). The displacement equations of motion of the sandwich plate then yield the frequency equation

$$\frac{r_\rho r_h^2}{r_\mu} (4 + 3r_\rho r_h) f^4 - 12 \left[\kappa_1 \frac{r_\rho r_h^2}{r_\mu} + \kappa_2 (1 + 3r_\rho r_h) \right] f^2 + 36 \kappa_1 \kappa_2 = 0, \quad (4.1.11)$$

where

$$r_\rho = \frac{\rho_2}{\rho_1}, \qquad r_h = \frac{h_2}{h_1}, \qquad r_\mu = \frac{\mu_2}{\mu_1}, \qquad f = \omega h_1 \sqrt{\frac{\rho_1}{\mu_1}}. \qquad (4.1.12)$$

The values of the shear factors κ_1 and κ_2 next are determined by matching the roots of f given by Eq. (4.1.11) with those given previously by Eq. (2.7.10a). For simplicity, we restrict ourselves to the lowest frequency of an ordinary sandwich plate, for which

$$r_h < 1, \qquad r_\rho > 1, \qquad r_\mu \gg 1 \tag{4.1.13}$$

and thus

$$\frac{r_\rho r_h^2}{r_\mu} \ll 1, \qquad \frac{r_\rho r_h^2}{r_\mu} f^2 \ll 1. \tag{4.1.14}$$

Under the conditions imposed by Eq. (4.1.14), Eq. (4.1.11) yields only the lowest root

$$f^2 = \frac{3\kappa_1}{1 + 3r_\rho r_h}. \tag{4.1.15}$$

Under the same conditions, the exact result in Eq. (2.7.10a) reduces to Eq. (2.7.14a):

$$f \tan f = \frac{1}{r_\rho r_h}. \tag{4.1.16}$$

The task before us is to determine κ_1 only so that f given by Eq. (4.1.15) matches the lowest f given by Eq. (4.1.16). The result clearly depends on the value of the product $r_\rho r_h$.

For $r_\rho r_h = 0$, the lowest root given by Eq. (4.1.16) is $f = \pi/2$, which, when substituted into Eq. (4.1.15), yields $\kappa_1 = \pi^2/12$, the value for a single-layered homogeneous plate. For sufficiently large $r_\rho r_h$, Eq. (4.1.16) becomes

$$f^2 = \frac{1}{r_\rho r_h},$$

and Eq. (4.1.15) gives $\kappa_1 = 1$. For $r_\rho r_h > 2$, numerical results reveal that κ_1 is essentially equal to 1, within an error not greater than one-half of a percent. Since $\kappa_1 = 1$ would reflect a perfect match between the sandwich plate and elasticity solutions, this agrees with our early finding (Yu 1959a,b) that the mode shapes predicted by the exact and approximate solutions are also similar. To summarize, we have found

$$\kappa_1 = \frac{\pi^2}{12} \qquad \text{for} \quad r_\rho r_h = 0 \tag{4.1.17}$$
$$\kappa_1 \approx 1 \qquad \text{for} \quad r_\rho r_h > 2.$$

We shall not attempt to determine the other shear factor κ_2, which does not serve a real purpose, for the following reason. Exact results in Figure 2.7.1 given by the elasticity solution indicate that the frequency of the lowest antisymmetric thickness-stretch mode may lie between the frequencies of the two lowest thickness-shear modes. Since a thickness-stretch mode has not been included in our sandwich plate theory, validity of the theory becomes questionable for the higher thickness-shear mode.

4.2 Simplified Refined Equations for Flexure of a Sandwich Plate with Membrane Facings

A system of simplified refined equations for flexure of a sandwich plate is derived next by considering the face layers as membranes (Yu 1960c). Although the transverse shear effect in the face layers is no longer considered, the effect in the core is still included. The resulting equations are therefore still refined equations. They represent a special case of the more general refined equations for flexure given in Section 4.1.

4.2.1 Variational Equation of Motion

To include only the transverse shear effect in the core of the sandwich, we now assume the three-dimensional displacements in the form

$$u_{x1} = z\psi, \qquad u_{x2}, u_{x3} = \pm h_1 \psi$$
$$u_{y1} = z\phi, \qquad u_{y2}, u_{y3} = \pm h_1 \phi \qquad (4.2.1)$$
$$u_{z1} = u_{z2} = u_{z3} = w.$$

This is clearly a special case of Eqs. (4.1.1) in which

$$\psi_1 = \psi, \qquad \phi_1 = \phi, \qquad \psi_2 = \phi_2 = 0. \qquad (4.2.2)$$

Replacing Eqs. (4.1.1) with (4.2.1) or (4.2.2) and following the same procedure used before, we find

$$\int_{t_0}^{t_1} dt \int \int_A \left\{ \left[\frac{\partial}{\partial x}\{M_{x1} + h_1(N_{x3} - N_{x2})\} + \frac{\partial}{\partial y}\{M_{yx1} + h_1(N_{yx3} - N_{yx2})\} \right. \right.$$
$$- Q_{x1} + h_1(p_x^+ - p_x^-) + f_{x1}^{(1)} + h_1(f_{x3}^{(0)} - f_{x2}^{(0)})$$
$$\left. - \left(\frac{2}{3}\rho_1 h_1^3 + 2\rho_2 h_2 h_1^2\right) \ddot{\psi} \right] \delta\psi$$

$$+ \left[\frac{\partial}{\partial x}\{M_{xy1} + h_1(N_{xy3} - N_{xy2})\} \right.$$
$$+ \frac{\partial}{\partial y}\{M_{y1} + h_1(N_{y3} - N_{y2})\}$$
$$- Q_{y1} + h_1(p_y^+ - p_y^-) + f_{y1}^{(1)} + h_1(f_{y3}^{(0)} - f_{y2}^{(0)})$$
$$\left. - \left(\frac{2}{3}\rho_1 h_1^3 + 2\rho_2 h_2 h_1^2\right) \ddot{\phi} \right] \delta\phi$$

$$+ \left[\frac{\partial Q_{x1}}{\partial x} + \frac{\partial Q_{y1}}{\partial y} + p_z^+ - p_z^- \right.$$
$$\left. \left. + f_{z1}^{(0)} + f_{z2}^{(0)} + f_{z3}^{(0)} - 2(\rho_1 h_1 + \rho_2 h_2)\ddot{w} \right] \delta w \right\} dx\, dy$$

$$-\int_{t_0}^{t_1} dt \oint_{C_p} \left\{ [M_{n1} + h_1(N_{n3} - N_{n2}) - p_{n1}^{(1)} - h_1(p_{n3}^{(0)} - p_{n2}^{(0)})]\,\delta\psi_n \right.$$

$$+ [M_{ns1} + h_1(N_{ns3} - N_{ns2}) - p_{s1}^{(1)} - h_1(p_{s3}^{(0)} - p_{s2}^{(0)})]\,\delta\psi_s$$

$$\left. + [Q_{n1} - p_{z1}^{(0)} - p_{z2}^{(0)} - p_{z3}^{(0)}]\,\delta w \right\}\, ds = 0, \tag{4.2.3}$$

where other notations are the same as before.

From the surface and boundary integrals in Eq. (4.2.3), the stress equations of motion and boundary conditions for the simplified case can be written similarly as before. The stress equations of motion are

$$\frac{\partial}{\partial x}\{M_{x1} + h_1(N_{x3} - N_{x2})\} + \frac{\partial}{\partial y}\{M_{yx1} + h_1(N_{yx3} - N_{yx2})\}$$

$$- Q_{x1} + h_1(p_x^+ - p_x^-) + f_{x1}^{(1)} + h_1(f_{x3}^{(0)} - f_{x2}^{(0)})$$

$$- \left(\frac{2}{3}\rho_1 h_1^3 + 2\rho_2 h_2 h_1^2\right)\ddot{\psi} = 0$$

$$\frac{\partial}{\partial x}\{M_{xy1} + h_1(N_{xy3} - N_{xy2})\} + \frac{\partial}{\partial y}\{M_{y1} + h_1(N_{y3} - N_{y2})\}$$

$$- Q_{y1} + h_1(p_y^+ - p_y^-) + f_{y1}^{(1)} + h_1(f_{y3}^{(0)} - f_{y2}^{(0)}) \tag{4.2.4}$$

$$- \left(\frac{2}{3}\rho_1 h_1^3 + 2\rho_2 h_2 h_1^2\right)\ddot{\phi} = 0$$

$$\frac{\partial Q_{x1}}{\partial x} + \frac{\partial Q_{y1}}{\partial y} + p_z^+ + p_z^-$$

$$+ f_{z1}^{(0)} + f_{z2}^{(0)} + f_{z3}^{(0)} - 2(\rho_1 h_1 + \rho_2 h_2)\ddot{w} = 0,$$

and the boundary conditions are

$$M_{n1} + h_1(N_{n3} - N_{n2}) - p_{n1}^{(1)} - h_1(p_{n3}^{(0)} - p_{n2}^{(0)}) = 0 \qquad \text{or} \quad \psi_n = \text{prescribed}$$

$$M_{ns1} + h_1(N_{ns3} - N_{ns2}) - p_{s1}^{(1)} - h_1(p_{s3}^{(0)} - p_{s2}^{(0)}) = 0 \quad \text{or} \quad \psi_s = \text{prescribed}$$

$$Q_{n1} - p_{z1}^{(0)} - p_{z2}^{(0)} - p_{z3}^{(0)} = 0 \qquad\qquad\qquad \text{or} \quad w = \text{prescribed.} \tag{4.2.5}$$

Another set of simplified refined equations of the sandwich plate (Yu 1960c) has been derived by assuming the displacements in the form, instead of Eqs. (4.2.1),

$$u_{x1} = z\psi, \qquad u_{x2}, u_{x3} = \mp\left(h_1 + \frac{h_2}{2}\right)\psi$$

$$u_{y1} = z\phi, \qquad u_{y2}, u_{y3} = \mp\left(h_1 + \frac{h_2}{2}\right)\phi \tag{4.2.6}$$

$$u_{z1} = u_{z2} = u_{z3} = w.$$

This is to assume the core of the sandwich to be extended slightly so that it would be joined to the face layers at the middle planes $z = \mp(h_1 + h_2/2)$ of the latter instead of at the interfaces. A detailed discussion of the resulting equations may be found in the original publication.

4.2.2 Plate Stress–Displacement Relations

These also can be obtained by the same procedure as used in the preceding section. Now that the face layers are considered as membranes, the moments and shearing forces in these layers have simply disappeared. The stress–displacement relations for an isotropic sandwich plate are thus

$$M_{x1}=D_1\left(\frac{\partial\psi}{\partial x}+\nu_1\frac{\partial\phi}{\partial y}\right)$$

$$M_{y1}=D_1\left(\frac{\partial\phi}{\partial y}+\nu_1\frac{\partial\psi}{\partial x}\right)$$

$$M_{xy1}=\tfrac{1}{2}D_1(1-\nu_1)\left(\frac{\partial\psi}{\partial y}+\frac{\partial\phi}{\partial x}\right)$$

$$Q_{x1}=2\kappa_1 G_1 h_1\left(\psi+\frac{\partial w}{\partial x}\right)$$

$$Q_{y1}=2\kappa_1 G_1 h_1\left(\phi+\frac{\partial w}{\partial y}\right) \qquad\qquad (4.2.7)$$

$$N_{x3}=-N_{x2}=C_2 h_1\left(\frac{\partial\psi}{\partial x}+\nu_2\frac{\partial\phi}{\partial y}\right)$$

$$N_{y3}=-N_{y2}=C_2 h_1\left(\frac{\partial\phi}{\partial y}+\nu_2\frac{\partial\psi}{\partial x}\right)$$

$$N_{xy3}=-N_{xy2}=\tfrac{1}{2}C_2(1-\nu_2)h_1\left(\frac{\partial\psi}{\partial y}+\frac{\partial\phi}{\partial x}\right),$$

where κ_1 is the only shear factor that still remains.

The core of the sandwich can be orthotropic. In such a case, the moments and shearing forces have the form

$$M_{x1}=D_1\frac{\partial\psi}{\partial x}+D_2\frac{\partial\phi}{\partial y}$$

$$M_{y1}=D_2\frac{\partial\psi}{\partial x}+D_3\frac{\partial\phi}{\partial y}$$

$$M_{xy1}=D_4\left(\frac{\partial\psi}{\partial y}+\frac{\partial\phi}{\partial x}\right) \qquad\qquad (4.2.8)$$

$$Q_{x1}=\kappa_1 c_{55}\delta_1\left(\psi+\frac{\partial w}{\partial x}\right)$$

$$Q_{y1} = \kappa_1 c_{66} \delta_1 \left(\phi + \frac{\partial w}{\partial y} \right),$$

where

$$D_1 = \left(c_{11} - \frac{c_{13}^2}{c_{33}} \right) \frac{2h_1^3}{3}$$

$$D_2 = \left(c_{12} - \frac{c_{13}c_{23}}{c_{33}} \right) \frac{2h_1^3}{3}$$

$$D_3 = \left(c_{22} - \frac{c_{23}^2}{c_{33}} \right) \frac{2h_1^3}{3}$$

$$D_4 = c_{44} \frac{2h_1^3}{3},$$

and the stiffnesses c_{ij} are given by Eqs. (1.7.6).

4.2.3 *Displacement Equations of Motion*

By means of the plate stress–displacement relations that have just been obtained, the stress equations (4.2.4) are easily transformed into the displacement equations of motion of the sandwich plate. In the case of an isotropic sandwich plate, the displacement equations have the form

$$\left[\tfrac{1}{2} D_1(1 - \nu_1) + C_2 h_1^2(1 - \nu_2) \right] \nabla^2 \psi + \left[\tfrac{1}{2} D_1(1 + \nu_1) + C_2 h_1^2(1 + \nu_2) \right] \frac{\partial e}{\partial x}$$

$$- 2\kappa_1 G_1 h_1 \left(\psi + \frac{\partial w}{\partial x} \right) + h_1(p_x^+ - p_x^-) + f_{x1}^{(1)}$$

$$+ h_1 \left(f_{x3}^{(0)} - f_{x2}^{(0)} \right) - \left(\frac{2}{3} \rho_1 h_1^3 + 2\rho_2 h_2 h_1^2 \right) \ddot{\psi} = 0$$

$$\left[\tfrac{1}{2} D_1(1 - \nu_1) + C_2 h_1^2(1 - \nu_2) \right] \nabla^2 \phi + \left[\tfrac{1}{2} D_1(1 + \nu_1) + C_2 h_1^2(1 + \nu_2) \right] \frac{\partial e}{\partial y}$$

$$- 2\kappa_1 G_1 h_1 \left(\phi + \frac{\partial w}{\partial y} \right) + h_1(p_y^+ - p_y^-) + f_{y1}^{(1)} \qquad (4.2.9)$$

$$+ h_1 (f_{y3}^{(0)} - f_{y2}^{(0)}) - \left(\frac{2}{3} \rho_1 h_1^3 + 2\rho_2 h_2 h_1^2 \right) \phi = 0$$

$$2\kappa_1 G_1 h_1 (\nabla^2 w + e) + p_z^+ - p_z^-$$

$$+ f_{z1}^{(0)} + f_{z2}^{(0)} + f_{z3}^{(0)} - 2(\rho_1 h_1 + \rho_2 h_2) \ddot{w} = 0,$$

where

$$e = \frac{\partial \psi}{\partial x} + \frac{\partial \phi}{\partial y}.$$

4.2.4   Determination of Shear Factor

As was shown in the preceding section, only a single shear factor κ_1 was determined. In fact, this is the only shear factor that remains when face layers are taken as membranes. In the case of an isotropic ordinary sandwich plate, the situation becomes the same as discussed in the preceding section, and results for κ_1 are also the same.

4.2.5   Analogy Between Sandwich and Homogeneous Plates

It is interesting to note that Eqs. (4.2.9) for an isotropic sandwich plate with membrane facings are mathematically analogous to Eqs. (3.2.10) for a single-layered homogeneous plate in which the transverse shear effect is also included. The analogy also holds true for the corresponding boundary conditions. Solutions for a homogeneous plate derived from Eqs. (3.2.10) therefore may be transformed into those for a sandwich plate directly.

4.3   Classical Equations for Flexure of a Sandwich Plate

To further reduce the flexural equations of a sandwich plate to the classical type, the transverse shear deformation in the core of the sandwich also is suppressed by letting

$$\psi = -\frac{\partial w}{\partial x}, \qquad \phi = -\frac{\partial w}{\partial y}. \tag{4.3.1}$$

In addition, the transverse shearing forces Q_{x1} and Q_{y1} must be eliminated from the three stress equations of motion (4.2.4). This is accomplished by differentiating the first of these equations with respect to x and the second with respect to y and adding the results to the third. The result is the following single stress equation of motion:

$$\frac{\partial^2}{\partial x^2}\{M_{x1} + h_1(N_{x3} - N_{x2})\} + 2\frac{\partial^2}{\partial x\,\partial y}\{M_{xy1} + h_1(N_{xy3} - N_{xy2})\}$$

$$+ \frac{\partial^2}{\partial y^2}\{M_{y1} + h_1(N_{y3} - N_{y2})\} + p_z^+ + p_z^- + f_{z1}^{(0)} + f_{z2}^{(0)} + f_{z3}^{(0)}$$

$$+ \frac{\partial}{\partial x}\{h_1(p_x^+ - p_x^-) + f_{x1}^{(0)} + h_1(f_{x3}^{(0)} - f_{x2}^{(0)})\}$$

$$+ \frac{\partial}{\partial y}\{h_1(p_y^+ - p_y^-) + f_{y1}^{(1)} + h_1(f_{y3}^{(0)} - f_{y2}^{(0)})\}$$

$$- 2(\rho_1 h_1 + \rho_2 h_2)\ddot{w} + \left(\frac{2}{3}\rho_1 h_1^3 + 2\rho_2 h_2 h_1^2\right)\nabla^2\ddot{w} = 0. \tag{4.3.2}$$

Elimination of ψ and ϕ from the stress–displacement relations is similar. In the case of an isotropic sandwich plate, the moments and membrane forces in Eqs. (4.2.7) become

$$M_{x1} = -D_1 \left(\frac{\partial^2 w}{\partial x^2} + \nu_1 \frac{\partial^2 w}{\partial y^2} \right)$$

$$M_{y1} = -D_1 \left(\frac{\partial^2 w}{\partial y^2} + \nu_1 \frac{\partial^2 w}{\partial x^2} \right)$$

$$M_{xy1} = -D_1(1 - \nu_1) \frac{\partial^2 w}{\partial x \partial y}$$

$$N_{x3} = -N_{x2} = -C_2 h_1 \left(\frac{\partial^2 w}{\partial x^2} + \nu_2 \frac{\partial^2 w}{\partial y^2} \right) \tag{4.3.3}$$

$$N_{y3} = -N_{y2} = -C_2 h_1 \left(\frac{\partial^2 w}{\partial y^2} + \nu_2 \frac{\partial^2 w}{\partial x^2} \right)$$

$$N_{xy3} = -N_{xy2} = -C_2 h_1 (1 - \nu_2) \frac{\partial^2 w}{\partial x \partial y}.$$

The transverse shearing forces in Eqs. (4.2.7) no longer apply. For a classical sandwich plate the displacement equations of motion are obtained by substituting Eqs. (4.3.3) into (4.3.2). The boundary conditions are treated similarly.

4.4   Flexural Vibration of an Infinite Sandwich Plate: Useful Ranges of Sandwich Plate Equations

In this section, the three systems of flexural equations that have just been derived are applied to the vibration analysis of an infinite sandwich plate. A comparison is made first between the exact equations of linear elasticity and the general system of refined equations of a sandwich plate. Next, for an ordinary sandwich plate that has relatively thin but heavy and rigid face layers, comparison is made among all three systems of sandwich plate equations. Finally, the effects of transverse shear and rotatory inertia are singled out for special assessment. An important consideration is to determine the ranges of frequency and wavelength (or wave number) within which each system of equations can be applied accurately and effectively, if the system of equations is indeed valid.

4.4.1   Comparison Between Elasticity and Refined Sandwich Plate Equations

To establish the reliability of the general system of refined equations of a sandwich plate derived in Section 4.1, these are compared with linear elasticity equations by investigating plane–strain modes of free vibration in an infinite sandwich plate. Based on the elasticity equations, a study has already been initiated in Section 2.6 (Yu 1960a). For plane–strain flexural modes, the frequency equation is given by Eq. (2.6.6), by means of which the circular frequency ω can be calculated for a given sandwich when a value of the wave number ξ is assigned.

Based on the general refined sandwich plate equations in Section 4.1, the displacement equations of motion are obtainable by substituting the plate stress–displacement relations from Eqs. (4.1.7) through (4.1.10) into the stress equations of motion (4.1.4). For plane–strain modes of free vibration, we ignore the surface and body forces in the displacement equations and choose to take the plate displacements in the form

$$w = W \sin \xi x \, e^{i\omega t}, \quad \psi_1 = \Psi_1 \cos \xi x \, e^{i\omega t}, \quad \psi_2 = \Psi_2 \cos \xi x \, e^{i\omega t}$$
$$\phi_1 = \phi_2 = 0. \tag{4.4.1}$$

The frequency equation then is obtained by setting equal to zero the determinant of the coefficients of the amplitudes W, Ψ_1, and Ψ_2. Although the result will not be written out in full here, it may be found in the original publication (Yu 1960a) and can be put in the form of the following cubic equation:

$$A_{30}\Omega^3 - (A_{20} + A_{22}\lambda^2)\Omega^2 + (A_{10-} + A_{12}\lambda^2 + A_{14}\lambda^4)\Omega - (A_{04}\lambda^4 + A_{06}\lambda^6) = 0, \tag{4.4.2}$$

where all notations are now dimensionless. In particular, the coefficients A_{ij} may be calculated for a sandwich when the ratios r_h, r_ρ, r_μ, $r_1 = 2/(1 - \nu_1)$ and $r_2 = 2r_\mu/(1 - \nu_2)$ are given. Then, instead of calculating the frequency ω versus the wave number ξ as from Eq. (2.6.6), the dimensionless frequency parameter $\Omega = (\rho_1/\mu_1)\omega^2 h^2$ can be calculated versus the dimensionless wave number parameter $\lambda = 2\pi h/L = \xi h$ from Eq. (4.4.2), where L is the wavelength. Three values of Ω are given by Eq. (4.4.2) for an assigned value of λ, and three branches of the frequency spectrum can be plotted accordingly. The lowest of these is the flexural branch, for which w is predominant. The higher two are thickness-shear branches, for which ψ_1 and ψ_2 are predominant.

Numerical examples have been given for $r_h = 1/10$ and $1/2$, and for $r_\rho = 34.4$, $r_\mu = 1683$, $r_1 = 2.2$, and $r_2 = 4790$, with the shear factor determined to be $\kappa_1 = 0.997$. Results have shown excellent agreement between the predictions of the elasticity equations and those of the general refined sandwich plate equations for both values of r_h (Yu 1960a).

4.4.2 Comparison Among Sandwich Plate Equations for an Ordinary Sandwich Plate

We recall that the general refined plate equations of the sandwich can be applied to any symmetrically constructed three-layered plate, with arbitrary values of the various ratios. The frequency equation (4.4.2) is similarly applicable to such an arbitrary three-layered plate. We shall next concentrate on an ordinary sandwich that has relatively thin but heavy and rigid face layers, for which the ratios r_h, r_ρ, and r_μ are usually smaller, greater, and much greater, respectively, than unity. For the parameters λ and Ω on the order of unity, we then have

$$\frac{r_\rho r_h^2}{r_\mu}, \quad \frac{r_\rho r_h^2}{r_\mu}\Omega, \quad r_h^2, \quad r_h^2\lambda^2 \ll 1. \tag{4.4.3}$$

Under these conditions, Eq. (4.4.2) reduces to the simpler form

$$B_{20}\Omega^2 - (B_{10} + B_{12}\lambda^2)\Omega + B_{04}\lambda^4 + B_{06}\lambda^6 = 0 \qquad (4.4.4a)$$

or, when written out in full (Yu 1960a),

$$\frac{1}{\kappa_1}(1 + r_\rho r_h)\left(\frac{1}{3} + r_\rho r_h\right)\Omega^2 - \left\{(1 + r_h)^2(1 + r_\rho r_h)\right.$$
$$+ \left[\frac{1}{3} + r_\rho r_h(1 + r_h) + \frac{1}{\kappa_1}(1 + r_\rho r_h)\left(\frac{r_1}{3} + r_2 r_h\right)\right]\lambda^2\bigg\}\Omega$$
$$+ \left[\frac{r_1}{3} + r_2 r_h(1 + r_h)\right]\lambda^4 + \frac{4r_1 + 3r_2 r_h}{36\kappa_1(1 + r_h)^2}r_2 r_h\lambda^6 = 0. \qquad (4.4.4b)$$

The reduced frequency equation now yields only two frequencies: one still for the flexural branch and the other for the lowest thickness-shear branch.

For an ordinary sandwich, the simplified refined plate equations are derived in Section 4.2 (Yu 1960c). The displacement equations of motion for an isotropic sandwich are given by Eqs. (4.2.9). For plane–strain modes of free vibration, we again ignore the external forces and now take

$$w = W \sin \xi x \, e^{i\omega t}, \qquad \psi = \Psi \cos \xi x \, e^{i\omega t}, \qquad \phi = 0.$$

The frequency equation derived from Eqs. (4.2.9) then has the form

$$B_{20}\Omega^2 - (B_{10} + B'_{12}\lambda^2)\Omega + B'_{04}\lambda^4 = 0 \qquad (4.4.5a)$$

or, when written out in full,

$$\frac{1}{\kappa_1}(1 + r_\rho r_h)\left(\frac{1}{3} + r_\rho r_h\right)\Omega^2 - \left\{(1 + r_h)^2(1 + r_\rho r_h)\right.$$
$$+ \left[\frac{1}{3} + r_\rho r_h + \frac{1}{\kappa_1}(1 + r_\rho r_h)\left(\frac{r_1}{3} + r_2 r_h\right)\right]\lambda^2\bigg\}\Omega$$
$$+ \left[\frac{r_1}{3} + r_2 r_h\right]\lambda^4 = 0. \qquad (4.4.5b)$$

Similar to Eq. (4.4.4b), this still yields two frequencies for a given sandwich. But there are differences between the two frequency equations. While the coefficients B_{20} and B_{10} are the same in both equations, the original coefficients B_{12} and B_{04} have been replaced by B'_{12} and B'_{04}, respectively, and the original B_{06} term has disappeared altogether. For such differences between the two frequency equations to be negligibly small, we find that the following conditions must be fulfilled:

$$r_h \ll 1, \qquad \frac{r_2 r_h^3}{12\kappa_1(1 + r_h)^2}\lambda^2 \ll 1. \qquad (4.4.6)$$

Table 4.4.1 $(r_h = 1/10)$

λ	$[r_2 r_h^3/12\kappa_1(1+r_h)^2]\lambda^2$
0.0	0
0.1	0.00330
1.0	0.330

The first of these is a rather restrictive condition requring that the face layers be very thin. The second condition is associated with the B_{06} term and involves both flexural and extensional rigidities of the face layers.

In the two numerical examples considered earlier in this section, the thickness ratio r_h is taken equal to 1/10 and 1/2, respectively. As an ordinary sandwich, we now consider only the thickness ratio $r_h = 1/10$, with other ratios and the shear factor remaining the same as before. While this thickness ratio does not quite fully satisfy the first of the two conditions in the inequalities (4.4.6), the second of these is very well satisfied for $\lambda = 0.1$ and not badly satisfied even for $\lambda = 1.0$, as is shown in Table 4.4.1.

Numerical results are calculated for the same ordinary sandwich with $r_h = 1/10$ by means of the frequency equations (4.4.4b) and (4.4.5b) derived from the general and simplified refined sandwich plate equations, respectively. These produce both the flexural and thickness-shear branches of the sandwich. The flexural branch also is calculated from the classical sandwich plate equations in Section 4.3, although the thickness-shear branch is no longer provided. The results are given in Table 4.4.2 and show very good agreement between predictions of the general and simplified refined equations for the flexural branch for $\lambda = 0.1$, satisfactory agreement even for $\lambda = 1.0$, and excellent agreement for the thickness-shear branch for the entire range of λ. In contrast, frequencies given by the classical equations for the flexural branch are drastically different and clearly unacceptable. The use of classical equations for ordinary sandwiches thus is not recommended.

Table 4.4.2 $(r_h = 1/10)$

	Lower $\sqrt{\Omega}$ (Flexural Branch)		
λ	General Refined Equations	Simplified Refined Equations	Classical Equations
0.0	0	0	0
0.1	0.0445	0.0424	0.1101
1.0	0.567	0.474	8.044

	Higher $\sqrt{\Omega}$ (Thickness-Shear Branch)	
λ	General Refined Equations	Simplified Refined Equations
0.0	0	0
0.1	1.239	1.240
1.0	11.26	11.30

4.4.3 Assessment of Effects of Transverse Shear and Rotatory Inertia

Both the general and simplified refined equations include the effects of transverse shear and rotatory inertia. Since they yield essentially similar results for the two lowest frequencies of an infinite ordinary sandwich plate, we choose to use the simplified refined equations to assess the importance of these effects. The assessment is carried out by means of the same procedure used for a homogeneous plate in Section 3.5. Thus, for plane–strain modes of free vibration, the displacement equations of motion (4.2.9) reduce to

$$D_{12}\frac{\partial^2 \psi}{\partial x^2} - 2\kappa_1 G_1 h_1 \left(\psi + \frac{\partial w}{\partial x}\right) - \rho_{12}^{(2)}\ddot{\psi} = 0$$

$$2\kappa_1 G_1 h_1 \left(\frac{\partial \psi}{\partial x} + \frac{\partial^2 w}{\partial x^2}\right) - \rho_{12}^{(0)}\ddot{w} = 0,$$

(4.4.7)

where

$$D_{12} = D_1 + 2C_2 h_1^2 = \frac{E_1}{1 - v_1^2}\frac{2h_1^3}{3} + \frac{E_2}{1 - v_2^2}(2h_1^2 h_2)$$

$$\rho_{12}^{(0)} = 2(\rho_1 h_1 + \rho_2 h_2), \qquad \rho_{12}^{(2)} = \rho_1\left(\frac{2h_1^3}{3}\right) + \rho_2(2h_1^2 h_2).$$

These are not only reducible to Eqs. (3.5.4) for a homogeneous plate as a special case, but also analogous to the latter equations. By extending the analogy further, therefore, the frequency equation for an infinite ordinary sandwich plate can be written in a form similar to Eq. (3.5.6):

$$\frac{\rho_{12}^{(2)}}{2\kappa_1 G_1 h_1}\omega^4 - \left[1 + \xi^2\frac{D_{12}}{2\kappa_1 G_1 h_1} + \xi^2\frac{\rho_{12}^{(2)}}{\rho_{12}^{(0)}}\right]\omega^2$$

$$+ \frac{D_{12}}{\rho_{12}^{(0)}}\xi^4 = 0,$$

(4.4.8)

which is exactly the same as Eq. (4.4.5b) in spite of the different notations used. Similar to Eq. (3.5.6) for a homogeneous plate, Eq. (4.4.8) can be used readily to assess the importance of the effects of transverse shear and rotatory inertia in a sandwich plate. Thus, for the transverse shear effect to be negligible, we find from the first two terms in Eq. (4.4.8) the following required conditions:

$$\frac{\rho_{12}^{(2)}}{2\kappa_1 G_1 h_1}\omega^2 = \frac{1}{\kappa_1}\frac{(1 + 3r_\rho r_h)}{3(1 + r_h)^2}\Omega \ll 1$$

$$\frac{D_{12}}{2\kappa_1 G_1 h_1}\xi^2 = \frac{1}{3\kappa_1}(r_1 + 3r_2 r_h)\lambda^2 \ll 1.$$

(4.4.9)

Similarly, for the rotatory inertia to be negligible, we find again from the second term in Eq. (4.4.8) another required condition:

$$\frac{\rho_{12}^{(2)}}{\rho_{12}^{(0)}}\xi^2 = \frac{1}{3}\frac{1+3r_\rho r_h}{1+r_\rho r_h}\lambda^2 \ll 1. \tag{4.4.10}$$

Among the three required conditions, the second is the most stringent for an ordinary sandwich because of the large value of r_2 involved. Consider the same numerical example discussed earlier. The second condition yields a value of $(r_1 + 3r_2 r_h)\lambda^2/3x_1$ equal to 4.81 even for $\lambda = 0.1$. Clearly, the transverse shear effect is not negligible.

The situation is further demonstrated in Figure 4.4.1, which shows the flexural and thickness-shear branches of the frequency spectrum for the same infinite sandwich plate. Comparing this with Figure 3.5.2 for a homogeneous plate, we also see clearly the difference between the two types of plates. While the classical results for a homogeneous plate are close to the corresponding refined and exact results for a substantial frequency range at the lower end, no such closeness is found in the sandwich plate starting from the origin. Numerically, the classical theory is easily applicable to a homogeneous plate up to several thousand hertz, but it may well lead to erroneous predictions for a sandwich plate at a level of, say, only 100 hertz. Thus we again reach the conclusion that the use of classical equations is simply not acceptable for ordinary sandwiches.

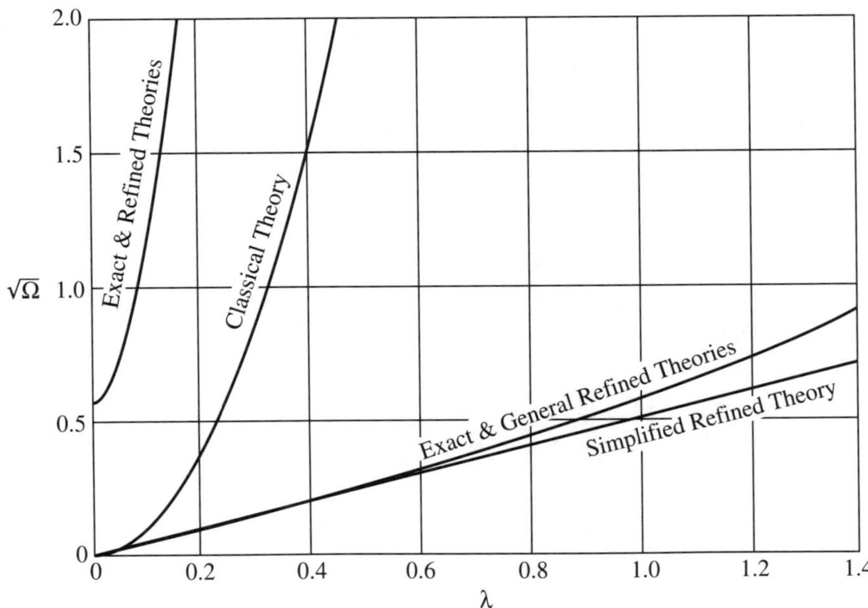

Fig. 4.4.1 Lowest two branches (flexural and thickness-shear) of frequency spectrum of flexural vibration of an infinite sandwich plate for $r_h = 1/10$.

4.5   Extensional Vibration of an Infinite Sandwich Plate Based on Classical Equations

Based on the exact equations of linear elasticity, the extensional vibration of an infinite homogeneous plate has been discussed in Chapter 2. Based on the classical and refined plate equations for extension, this has further been examined in Chapter 3. As was noted earlier, the refined equations of Kane and Mindlin (1956) are only good for a Poisson's ratio smaller than 1/3. For a ratio greater than 1/3, further refinement of plate equations must be made, as is shown by Mindlin and Medick (1959).

In this section, we treat the extensional vibration of an infinite sandwich plate. Similar complication arises as in the case of a homogeneous plate. Fortunately, for an ordinary sandwich plate with relatively thin but heavy and rigid face layers, the lowest thickness-stretch frequency is usually already at a very high level. For a frequency range below this level, classical equations for extension of a sandwich plate are usually accurate enough for ordinary engineering purposes. Thus, we shall cover here only the derivation of the classical equations and their application to the vibration analysis of an infinite sandwich plate. Discussion of refined equations for extension of a sandwich plate may be found in our early and more comprehensive work (Yu 1962b).

4.5.1   Variational Equation of Motion

To derive the classical equations for extension of a sandwich plate, the three-dimensional displacements are taken in the form

$$u_{xi}(x, y, z, t) = u(x, y, t), \ u_{yi}(x, y, z, t) = v(x, y, t), \ u_{zi}(x, y, z, t) = 0$$
$$(i = 1, 2, 3), \tag{4.5.1}$$

where u and v are the two-dimensional plate displacements. Substituting Eqs. (4.5.1) into the variational equation of motion (1.4.9) and carrying out integration with respect to z over the total thickness of the sandwich plate as before, we find

$$\int_{t_0}^{t_1} dt \int_A \int \left\{ \left[\frac{\partial}{\partial x}(N_{x1} + N_{x2} + N_{x3}) + \frac{\partial}{\partial y}(N_{xy1} + N_{xy2} + N_{xy3}) \right. \right.$$

$$\left. + P_x^{(0)} + f_{x1}^{(0)} + f_{x2}^{(0)} + f_{x3}^{(0)} - 2(\rho_1 h_1 + \rho_2 h_2)\ddot{u} \right] \delta u$$

$$+ \left[\frac{\partial}{\partial x}(N_{xy1} + N_{xy2} + N_{xy3}) + \frac{\partial}{\partial y}(N_{y1} + N_{y2} + N_{y3}) \right.$$

$$+ P_y^{(0)} + f_{y1}^{(0)} + f_{y2}^{(0)} + f_{y3}^{(0)} - 2(\rho_1 h_1 + \rho_2 h_2)\ddot{v}\Big]\delta v\Big\} dx\, dy$$

$$- \int_{t_0}^{t_1} dt \oint_C \Big\{ \Big[N_{n1} + N_{n2} + N_{n3} - p_{n1}^{(0)} - p_{n2}^{(0)} - p_{n3}^{(0)} \Big]\delta u_n$$

$$+ \Big[N_{ns1} + N_{ns2} + N_{ns3} - p_{s1}^{(0)} - p_{s2}^{(0)} - p_{s3}^{(0)} \Big]\delta u_s \Big\}\, ds = 0,$$

$$(4.5.2)$$

where the notations are similar to those in Section 3.3:

$$(N_{xi}, N_{yi}, N_{xyi}) = \int_{b_i}^{a_i} (\sigma_{xxi}, \sigma_{yyi}, \sigma_{xyi})\, dz$$

$$P_x^{(0)} = p_x^+ + p_x^-, \qquad P_y^{(0)} = p_y^+ + p_y^-$$

$$f_{xi}^{(0)} = \int_{b_i}^{a_i} f_{xi}\, dz, \qquad f_{yi}^{(0)} = \int_{b_i}^{a_i} f_{yi}\, dz$$

$$(p_{ni}^{(0)}, p_{si}^{(0)}) = \int_{b_i}^{a_i} (p_{ni}, p_{si})\, dz.$$

Equation (4.5.2) is the variational equation of motion for extension of the sandwich plate, from which the stress equations of motion and boundary conditions may be written in the usual manner.

4.5.2 Displacement Equations of Motion

The plate stress–displacement relations are similar to those given by Eqs. (3.3.7) and (3.3.8):

$$N_{xi} = C_i \left(\frac{\partial u}{\partial x} + v_i \frac{\partial v}{\partial y} \right)$$

$$N_{yi} = C_i \left(\frac{\partial v}{\partial y} + v_i \frac{\partial u}{\partial x} \right) \qquad (i = 1, 2, 3), \qquad (4.5.3)$$

$$N_{xyi} = \frac{1}{2} C_i (1 - v_i) \left(\frac{\partial u}{\partial y} + \frac{\partial v}{\partial x} \right)$$

where

$$C_i = \frac{E_i \delta_i}{1 - v_i^2}$$

is the extensional rigidity of the ith layer.

By substituting Eqs. (4.5.3) into the stress equations of motion written from Eq. (4.5.2), the displacement equations of motion of the sandwich plate are obtained readily:

$$\left[\frac{E_1 h_1}{1+\nu_1} + \frac{E_2 h_2}{1+\nu_2}\right]\nabla^2 u + \left[\frac{E_1 h_1}{1+\nu_1} + \frac{E_2 h_2}{1+\nu_2}\right]\frac{\partial e}{\partial x}$$
$$+ P_x^{(0)} + f_{x1}^{(0)} + f_{x2}^{(0)} + f_{x3}^{(0)} - 2(\rho_1 h_1 + \rho_2 h_2)\ddot{u} = 0$$
$$\left[\frac{E_1 h_1}{1+\nu_1} + \frac{E_2 h_2}{1+\nu_2}\right]\nabla^2 v + \left[\frac{E_1 h_1}{1+\nu_1} + \frac{E_2 h_2}{1+\nu_2}\right]\frac{\partial e}{\partial y}$$
$$+ P_y^{(0)} + f_{y1}^{(0)} + f_{y2}^{(0)} + f_{y3}^{(0)} - 2(\rho_1 h_1 + \rho_2 h_2)\ddot{v} = 0$$

$$(4.5.4)$$

where

$$e = \frac{\partial u}{\partial x} + \frac{\partial v}{\partial y}.$$

4.5.3   Extensional Vibration of an Infinite Sandwich Plate

For extensional modes of free vibration of an infinite sandwich plate, the external load terms in Eqs. (4.5.4) are not needed; and for plane strain modes, we choose to take

$$u = U \sin \frac{\lambda x}{h} e^{i\omega t}, \qquad v = 0.$$

The frequency then is given by

$$\Omega = \lambda^2 \frac{r_1 + r_2 r_h}{1 + r_\rho r_h}, \tag{4.5.5}$$

where notations for the ratios are the same as before. This is clearly a simple linear relation between the circular frequency ω and the wave number parameter λ, as in the case of a homogeneous plate and, in fact, is redeucible to the latter as a special case.

It will be useful to be able to calculate the lowest simple thickness-stretch frequency of an infinite sandwich plate. This is higher than the lowest simple thickness-shear frequency and can be determined only from the more refined equations of a sandwich plate (Yu 1962b). The result gives

$$\Omega = \frac{3\kappa_1 r_1 (1 + r_h)^2}{1 + 3 r_\rho r_h}, \tag{4.5.6}$$

where the correction factor κ_1 may be taken equal to 1 for simplicity, as in the flexural case.

As a numerical example, we consider an ordinary sandwich plate with the same ratios used earlier (Yu 1960a):

$$r_h = 0.1, \qquad r_\rho = 34.4, \qquad r_\mu = 1683, \qquad r_1 = 2.2, \qquad r_2 = 4790.$$

From Eq. (4.5.6) we find

$$\Omega = 0.712.$$

For a sandwich plate with a total thickness equal to 0.3 inch, this corresponds to a very high frequency of 15,600 hertz. For a range up to this level, extensional frequencies of the sandwich can be predicted accurately by means of the simple linear relation in Eq. (4.5.5).

References

Bert, C.W. (1975) Analysis of Plates; Analysis of Shells. In: *Composite Materials*, Vol. 7, edited by C.C. Chamis, pp. 149–258. Academic Press, New York.

Habip, L.M. (1965) A Survey of Modern Developments in the Analysis of Sandwich Structures. *Applied Mechanics Reviews*, Vol. 18, pp. 93–98.

Kane, T.R. and R.D. Mindlin. (1956) High-Frequency Extensional Vibrations of Plates. *Journal of Applied Mechanics*, Vol. 23, pp. 277–283.

Knoell, A.C. and E.Y. Robinson. (1975) Analysis of Truss, Beam, Frame, and Membrane Composites. In: *Composite Materials*, Vol. 7, edited by C.C. Chamis, Academic Press, New York.

Mindlin, R.D. (1951) Influence of Rotatory Inertia and Shear on Flexural Motions of Isotropic, Elastic Plates. *Journal of Applied Mechanics*, Vol. 18, pp. 31–38.

Mindlin, R.D. and M.A. Medick. (1959) Extensional Vibrations of Elastic Plates. *Journal of Applied Mechanics*, Vol. 26, pp. 561–569.

Nicholas, T. and R.A. Heller. (1967) Determination of the Complex Shear Modulus of a Filled Elastomer from a Vibrating Sandwich Beam. *Experimental Mechanics*, Vol. 7, pp. 110–116.

Rao, Y.V.K.S. and B.C. Nakra. (1974) Vibrations of Unsymmetrical Sandwich Beams and Plates with Viscoelastic Cores. *Journal of Sound and Vibrations*, Vol. 34, pp. 309–326.

Toledano, A. and H. Murakami. (1987) A Composite Plate Theory for Arbitrary Laminate Configurations. *Journal of Applied Mechanics*, Vol. 54, pp. 181–189.

Yu, Y.Y. (1959a) A New Theory of Elastic Sandwich Plates—One-Dimensional Case. *Journal of Applied Mechanics*, Vol. 26, pp. 415–421.

Yu, Y.Y. (1959b) Simple Thickness-Shear Modes of Vibration of Infinite Sandwich Plates. *Journal of Applied Mechanics*, Vol. 26, pp. 679–681.

Yu, Y.Y. (1960a) Flexural Vibrations of Elastic Sandwich Plates. *Journal of Aero/Space Sciences*, Vol. 27, pp. 272–283.

Yu, Y.Y. (1960b) Forced Flexural Vibrations of Elastic Sandwich Plates in Plane Strain. *Journal of Applied Mechanics*, Vol. 27, pp. 535–540.

Yu, Y.Y. (1960c) Simplified Vibration Analysis of Elastic Sandwich Plates. *Journal of Aero/Space Sciences*, Vol. 27, pp. 894–900.

Yu, Y.Y. (1962a) Damping of Flexural Vibrations of Sandwich Plates. *Journal of Aero/Space Sciences*, Vol. 29, pp. 790–803.

Yu, Y.Y. (1962b) Extensional Vibrations of Elastic Sandwich Plates. In: *Proceedings of the Fourth U.S. National Congress of Applied Mechanics*, pp. 441–447. ASME, New York.

Yu, Y.Y. (1989) Dynamics and Vibration of Layered Plates and Shells—A Perspective from Sandwiches to Laminated Composites. Bound volume for the 30th AIAA/ASME/ASCE/AHS/ASC Structures, Structural Dynamics, and Materials Conference, pp. 2236–2245.

Yu, Y.Y. (1992) Dynamics and Vibration of Layered Plates and Shells—A Perspective from Sandwiches to Laminated Composites (Russian translation). In: *Mechanics of Composite Materials, Vol. 2, Construction of Composites, Proceedings of the First Soviet– American Symposium*, pp. 99-108. Riga, Latvia.

5

Linear Modeling of Laminated Composite Plates

Laminated composite plates are often simply called *laminates*. In this chapter, linear equations of motion of laminates are derived and applied to vibration analysis. In general, laminates are multiple-layered and anisotropic, and thus, flexure and extension can be coupled in a laminate even in the linear case.

Many books on laminated composites are available, such as those by Jones (1975), Tsai and Hahn (1980), Vinson and Sierakowski (1987), Whitney (1987), and Agarwal and Broutman (1990). Early works on flexural vibrations of anisotropic plates were reported in the books by Lekhnitskii (1950) and Hearmon (1961), and the results are applicable to laminated composites. Transverse shear effect was included in linear equations of laminates by Whitney and Pagano (1970). Flexural vibrations of laminates have since been covered in many other books, papers, and reviews, as were cited in our aforementioned recent survey that has appeared both in English (Yu 1989) and in Russian (Yu 1992a).

In a recent study of linear equations and vibrations of laminates (Yu 1992b, 1995), one of the important conclusions concerns symmetric laminates, for which the equations become completely uncoupled, both statically and dynamically, into two subsystems for flexure and extension. Linear extensional vibrations of laminates were investigated for the first time in the same work. It was also found that the refined flexural equations of symmetric laminates, including shear and rotatory inertia effects, have exactly the same form as Mindlin's equations of crystal plates, and that shear effect can be equally important in vibrations of laminates as in sandwiches. Dudek (1970) had noted the possible importance of shear effect in his experimental work on laminates.

In this chapter, linear dynamical modeling of laminates is treated. in Section 5.1, we derive linear classical equations of motion for an arbitrary laminate, from which uncoupled equations for flexure and extension of a symmetric laminate are deduced as special cases. In Section 5.2, the transverse shear effect is taken into account, and linear refined equations are derived. In Sections 5.3 and 5.4, we investigate linear flexural and extensional vibrations, respectively, of an infinite symmetric laminate, by means of which the useful ranges of the flexural and extensional equations are determined. Specifically, conditions are presented under which the effects of transverse shear and rotatory inertia may be neglected or must be considered. This helps in determining whether the classical or refined equations should be selected in an engineering analysis of laminates.

5.1   Classical Equations of a Laminated Composite Plate

We consider a laminated composite plate or laminate with the same overall dimensions as given before for a general plate in Figure 3.1.1. In particular, the total thickness of the plate or laminate is $2h$, although the laminate is composed of a number of plies. Even when the reinforcement in each ply is unidirectional, the laminate as a whole can be multidirectional. The laminate is usually not only multiple-layered but also anisotropic. We let the total number of plies in the laminate be designated by n and let the kth ply extend from $z = h_{k-1}$ to $z = h_k$.

5.1.1   Variational Equation of Motion

To combine flexure and extension, we take the three-dimensional displacements in the form, according to Eqs. (3.1.1) and (3.3.1),

$$u_x(x, y, z, t) = u(x, y, t) - z\frac{\partial w}{\partial x}$$

$$u_y(x, y, z, t) = v(x, y, t) - z\frac{\partial w}{\partial y} \qquad (5.1.1)$$

$$u_z(x, y, z, t) = w(x, y, t).$$

Following the procedure used before, we substitute Eqs. (5.1.1) into the linear variational equation of motion (1.4.9) and carry out integration with respect to z, now covering all plies of the laminate. There results (Yu 1992b)

$$\int_{t_0}^{t_1} dt \int\int_A \left\{ \left[\frac{\partial N_x}{\partial x} + \frac{\partial N_{yx}}{\partial y} + P_x^{(0)} + f_x^{(0)} - \rho^{(0)}\ddot{u} + \rho^{(1)}\frac{\partial \ddot{w}}{\partial x} \right] \delta u \right.$$

$$+ \left[\frac{\partial N_{xy}}{\partial x} + \frac{\partial N_y}{\partial y} + P_y^{(0)} + f_y^{(0)} - \rho^{(0)}\ddot{v} + \rho^{(1)}\frac{\partial \ddot{w}}{\partial y} \right] \delta v$$

$$+ \left[\frac{\partial^2 M_x}{\partial x^2} + 2\frac{\partial M_{xy}}{\partial x \partial y} + \frac{\partial^2 M_y}{\partial y^2} \right.$$

$$+ \frac{\partial P_x^{(1)}}{\partial x} + \frac{\partial P_y^{(1)}}{\partial y} + P_z^{(0)}$$

$$+ \frac{\partial f_x^{(1)}}{\partial x} + \frac{\partial f_y^{(1)}}{\partial y} + f_z^{(0)}$$

$$\left. - \rho^{(0)} \ddot{w} - \rho^{(1)} \left(\frac{\partial \ddot{u}}{\partial x} - \frac{\partial \ddot{v}}{\partial y} \right) + \rho^{(2)} \nabla^2 \ddot{w} \right] \delta w \Big\} \, dx \, dy$$

$$- \int_{t_0}^{t_1} dt \oint_C \Big\{ [N_n - p_n^{(0)}] \, \delta u_n + [N_{ns} - p_s^{(0)}] \, \delta u_s$$

$$+ \left[\frac{\partial M_n}{\partial n} + 2 \frac{\partial M_{ns}}{\partial s} + P_n^{(1)} + f_n^{(1)} - p_z^{(0)} \right.$$

$$\left. - \frac{\partial p_s^{(1)}}{\partial s} - \rho^{(1)} \ddot{u}_n + \rho^{(2)} \frac{\partial \ddot{w}}{\partial n} \right] \delta w$$

$$- [M_n - p_n^{(1)}] \, \delta \left(\frac{\partial w}{\partial n} \right) \Big\} \, ds = 0, \tag{5.1.2}$$

where

$$[N_x, N_y, N_{xy}, M_x, M_y, M_{xy}]$$

$$= \sum_{k=1}^{n} \int_{h_{k-1}}^{h_k} [\sigma_{xx(k)}, \sigma_{yy(k)}, \sigma_{xy(k)}, \sigma_{xx(k)}z, \sigma_{yy(k)}z, \sigma_{xy(k)}z] \, dz$$

$$P_x^{(0)} = p_x^+ + p_x^-, \quad P_y^{(0)} = \ldots, \quad\quad P_z^{(0)} = \ldots,$$

$$P_n^{(0)} = \ldots, \quad\quad P_s^{(0)} = \ldots,$$

$$P_x^{(1)} = h(p_x^+ - p_x^-), \quad P_y^{(1)} = \ldots,$$

$$P_n^{(1)} = \ldots, \quad\quad P_s^{(1)} = \ldots$$

$$[f_x^{(0)}, f_x^{(1)}, f_y^{(0)}, f_y^{(1)}, f_z^{(0)}] \tag{5.1.3}$$

$$= \sum_{k=1}^{n} \int_{h_{k-1}}^{h_k} [f_{x(k)}, f_{x(k)}z, f_{y(k)}, f_{y(k)}z, f_{z(k)}] \, dz$$

$$[\rho^{(0)}, \rho^{(1)}, \rho^{(2)}] = \sum_{k=1}^{n} \int_{h_{k-1}}^{h_k} \rho_{(k)}[1, z, z^2] \, dz$$

$$[p_x^{(0)}, p_x^{(1)}] = \sum_{k=1}^{n} \int_{h_{k-1}}^{h_k} p_x[1, z] \, dz, \quad \ldots.$$

The subscript k in parentheses is used to denote the kth ply. Equation (5.1.2) is the variational equation of motion for a laminated composite plate from which the stress equations of motion and boundary conditions can be written in the usual manner.

As in the case of a classical homogeneous plate treated in Section 3.1, the transverse shearing forces Q_x and Q_y do not appear in the final form of the variational equation for a laminate of the classical type. As in the classical theory of a single-layered plate, p_x^+, p_x^-, ... in Eqs. (5.1.2) and (5.1.3) still represent only the tractions at the top and bottom boundary planes of the laminate. The tractions between adjacent plies are actions and reactions to each other and become cancelled out. In contrast to the single-layered case, however, Eq. (5.1.2) contains more inertia terms than the usual flexural and extensional equations due to the presence of the coupling inertia $\rho^{(1)}$, which may not vanish for an anisotropic laminate. This provides dynamic coupling between flexure and extension. The coupling disappears for a symmetric laminate. To deduce from Eq. (5.1.2) the result for a single-layered isotropic plate, we simplify the stiffnesses in the usual manner and let

$$\rho^{(0)} = 2\rho h, \qquad \rho^{(1)} = 0, \qquad \rho^{(2)} = \frac{2\rho h^3}{3}. \tag{5.1.4}$$

Flexure and extension then become uncoupled, both statically and dynamically.

5.1.2   Stress–Strain–Displacement Relations

Each ply in the fiber-reinforced composite laminate is assumed to have a plane of elastic symmetry for which the general stress–strain relations are specified by Eqs. (1.7.3) and (1.7.5). In the linear case, the stresses are given by

$$\begin{bmatrix} \sigma_{xx} \\ \sigma_{yy} \\ \sigma_{zz} \\ \sigma_{yz} \\ \sigma_{zx} \\ \sigma_{xy} \end{bmatrix} = \begin{bmatrix} c_{11} & c_{12} & c_{13} & 0 & 0 & c_{16} \\ c_{21} & c_{22} & c_{23} & 0 & 0 & c_{26} \\ c_{31} & c_{32} & c_{33} & 0 & 0 & c_{36} \\ 0 & 0 & 0 & c_{44} & c_{45} & 0 \\ 0 & 0 & 0 & c_{54} & c_{55} & 0 \\ c_{61} & c_{62} & c_{63} & 0 & 0 & c_{66} \end{bmatrix} \begin{bmatrix} e_{xx} \\ e_{yy} \\ e_{zz} \\ 2e_{yz} \\ 2e_{zx} \\ 2e_{xy} \end{bmatrix}, \tag{5.1.5}$$

where c_{ij} are the stiffnesses and e_{ij} are the linear tensorial strains. In contracted notations introduced in Eqs. (1.7.1), these relations may be written in the compact form

$$\sigma_i = c_{ij}\, e_j,$$

where e_j are the linear engineering strains. A ply is usually treated as an orthotropic layer with material axes parallel and normal to the fiber direction.

The plies are assumed to be in a state of plane stress so that $\sigma_{zz} = \sigma_3$ is set equal to 0 and $e_{zz} = e_3$ is eliminated from Eqs. (5.1.5) by solving for e_3 and substituting the result back into the same equations; thus,

$$\sigma_i = Q_{i\alpha}\, e_\alpha \qquad (i = 1, 2, 3, 6;\ \alpha = 1, 2, 6),$$

where

$$Q_{i\alpha} = c_{i\alpha} - \frac{c_{i3}}{c_{33}} c_{3\alpha}$$

are the reduced stiffnesses. For the kth ply, denoted by the subscript (k), the relations for σ_{xx}, σ_{yy}, and σ_{xy} in Eqs. (5.1.5) now are written in the following form:

$$
\begin{bmatrix} \sigma_{xx(k)} \\ \sigma_{yy(k)} \\ \sigma_{xy(k)} \end{bmatrix} = \begin{bmatrix} Q_{11(k)} & Q_{12(k)} & Q_{16(k)} \\ Q_{21(k)} & Q_{22(k)} & Q_{26(k)} \\ Q_{16(k)} & Q_{26(k)} & Q_{66(k)} \end{bmatrix} \begin{bmatrix} e_{xx} \\ e_{yy} \\ 2e_{xy} \end{bmatrix}. \tag{5.1.6}
$$

These are needed in the calculation of the plate stresses or the resultant forces and moments N_x, N_y, N_{xy}, M_x, M_y, and M_{xy}. Since the transverse shearing forces Q_x and Q_y do not appear in the stress equations of motion in the classical plate theory, the relations for σ_{yz} and σ_{zx} in Eqs. (5.1.5) are not needed at this time.

Corresponding to the displacements assumed in Eqs. (5.1.1), the linear strains have the typical form

$$
e_{xx} = e_{xx}^{(0)} + z\, e_{xx}^{(1)},
$$

where the superscripts 0 and 1 in parentheses denote plate strains of the zeroth and first order, respectively. For a multidirectional laminate, the plate stresses are obtained by integrating the stresses in Eqs. (5.1.6) over the thickness of the laminate according to Eqs. (5.1.3). The results are the following plate stress–strain relations:

$$
\begin{bmatrix} N_x \\ N_y \\ N_{xy} \\ M_x \\ M_y \\ M_{xy} \end{bmatrix} = \begin{bmatrix} A_{11} & A_{12} & A_{16} & B_{11} & B_{12} & B_{16} \\ A_{12} & A_{22} & A_{26} & B_{12} & B_{22} & B_{26} \\ A_{16} & A_{26} & A_{66} & B_{16} & B_{26} & B_{66} \\ B_{11} & B_{12} & B_{16} & D_{11} & D_{12} & D_{16} \\ B_{12} & B_{22} & B_{26} & D_{12} & D_{22} & D_{26} \\ B_{16} & B_{26} & B_{66} & D_{16} & D_{26} & D_{66} \end{bmatrix} \begin{bmatrix} e_{xx}^{(0)} \\ e_{yy}^{(0)} \\ 2e_{xy}^{(0)} \\ e_{xx}^{(1)} \\ e_{yy}^{(1)} \\ 2e_{xy}^{(1)} \end{bmatrix}, \tag{5.1.7}
$$

where

$$
[A_{ij}, B_{ij}, D_{ij}] = \sum_{k=1}^{n} Q_{ij(k)} \left[h_k - h_{k-1}, \frac{h_k^2 - h_{k-1}^2}{2}, \frac{h_k^3 - h_{k-1}^3}{3} \right]
$$

are the extensional stiffnesses, coupling stiffnesses, and flexural stiffnesses, respectively.

The strain–displacement relations are simply, according to Eqs. (5.1.1),

$$
e_{xx} = \frac{\partial u}{\partial x} - z\frac{\partial^2 w}{\partial x^2}
$$

$$
e_{yy} = \frac{\partial v}{\partial y} - z\frac{\partial^2 w}{\partial y^2}
$$

$$
e_{xy} = \frac{1}{2}\left(\frac{\partial u}{\partial y} + \frac{\partial v}{\partial x} \right) - z\frac{\partial^2 w}{\partial x \partial y}.
$$

The plate strains are thus

$$
e_{xx}^{(0)} = \frac{\partial u}{\partial x}
$$

$$e_{yy}^{(0)} = \frac{\partial v}{\partial y}$$

$$e_{xy}^{(0)} = \frac{1}{2}\left(\frac{\partial u}{\partial y} + \frac{\partial v}{\partial x}\right)$$

$$e_{xx}^{(1)} = k_x = -\frac{\partial^2 w}{\partial x^2}$$

$$e_{yy}^{(1)} = k_y = -\frac{\partial^2 w}{\partial y^2}$$

$$e_{xy}^{(1)} = k_{xy} = -\frac{\partial^2 w}{\partial x \partial y},$$

$$(5.1.8)$$

where k_x, k_y, and k_{xy} are the usual curvatures and twist. Equations (5.1.8) are the plate strain–displacement relations. Substitution of Eqs. (5.1.8) into (5.1.7) yields the plate stress–displacement relations. Substitution of the results, in turn, into the stress equations of motion leads finally to the displacement equations of motion.

5.1.3   Symmetric Laminate

As was noted earlier, the coupling inertia $\rho^{(1)}$ vanishes for a single-layered isotropic plate. Indeed, it vanishes for any symmetric laminate, the single-layered isotropic plate being merely a special case. A more widely known characteristic of a symmetric laminate consists of the vanishing coupling stiffnesses B_{ij} in the static case. Thus, flexure and extension become completely uncoupled from each other for a symmetric laminate. The displacement equation of motion for flexure is then

$$D_{11}\frac{\partial^4 w}{\partial x^4} + 4D_{16}\frac{\partial^4 w}{\partial x^3 \partial y} + 2(D_{12} + 2D_{66})\frac{\partial^4 w}{\partial x^2 \partial y^2} + 4D_{26}\frac{\partial^4 w}{\partial x \partial y^3}$$

$$+ D_{22}\frac{\partial^4 w}{\partial y^4} - \frac{\partial P_x^{(1)}}{\partial x} - \frac{\partial P_y^{(1)}}{\partial y} - P_z^{(0)}$$

$$- \frac{\partial f_x^{(1)}}{\partial x} - \frac{\partial f_y^{(1)}}{\partial y} - f_z^{(0)} + \rho^{(0)}\ddot{w} - \rho^{(2)}\nabla^2 \ddot{w} = 0, \qquad (5.1.9)$$

and the displacement equations for extension are

$$A_{11}\frac{\partial^2 u}{\partial x^2} + 2A_{16}\frac{\partial^2 u}{\partial x \partial y} + A_{66}\frac{\partial^2 u}{\partial y^2} + A_{16}\frac{\partial^2 v}{\partial x^2}$$

$$+ (A_{12} + A_{66})\frac{\partial^2 v}{\partial x \partial y} + A_{26}\frac{\partial^2 v}{\partial y^2} + P_x^{(0)} + f_x^{(0)} - \rho^{(0)}\ddot{u} = 0$$

$$(5.1.10)$$

$$A_{16}\frac{\partial^2 u}{\partial x^2} + (A_{12} + A_{66})\frac{\partial^2 u}{\partial x \partial y} + A_{26}\frac{\partial^2 u}{\partial y^2} + A_{66}\frac{\partial^2 v}{\partial x^2}$$

$$+ 2A_{26}\frac{\partial^2 v}{\partial x \partial y} + A_{22}\frac{\partial^2 v}{\partial y^2} + P_y^{(0)} + f_y^{(0)} - \rho^{(0)}\ddot{v} = 0.$$

It is well known that we further have $A_{16} = A_{26} = D_{16} = D_{26} = 0$ if the symmetric laminate is orthotropic. We also note that, according to Eq. (5.1.9), flexural motion can be excited by surface tractions and body forces in the x- and y-directions, as well as in the z-direction.

5.2   Refined Equations of a Laminated Composite Plate

We now continue our treatment of a laminated composite plate by incorporating the effect of transverse shear (Yu 1992b). The results are similar to those previously given by Whitney and Pagano (1970).

5.2.1   Variational Equation of Motion

We now take the three-dimensional displacements in the form, according to Eqs. (3.2.1) and (3.3.1),

$$u_x(x, y, z, t) = u(x, y, t) + z\psi(x, y, t)$$
$$u_y(x, y, z, t) = v(x, y, t) + z\phi(x, y, t) \tag{5.2.1}$$
$$u_z(x, y, z, t) = w(x, y, t).$$

The variational equation of motion for the laminate is then

$$
\int_{t_0}^{t_1} dt \int \int_A \left\{ \left[\frac{\partial N_x}{\partial x} + \frac{\partial N_{yx}}{\partial y} + P_x^{(0)} + f_x^{(0)} - \rho^{(0)}\ddot{u} - \rho^{(1)}\ddot{\psi} \right] \delta u \right.
$$
$$
+ \left[\frac{\partial N_{xy}}{\partial x} + \frac{\partial N_y}{\partial y} + P_y^{(0)} + f_y^{(0)} - \rho^{(0)}\ddot{v} - \rho^{(1)}\ddot{\phi} \right] \delta v
$$
$$
+ \left[\frac{\partial M_x}{\partial x} + \frac{\partial M_{yx}}{\partial y} - Q_x + P_x^{(1)} + f_x^{(1)} - \rho^{(1)}\ddot{u} - \rho^{(2)}\ddot{\psi} \right] \delta \psi
$$
$$
+ \left[\frac{\partial M_{xy}}{\partial x} + \frac{\partial M_y}{\partial y} - Q_y + P_y^{(1)} + f_y^{(1)} - \rho^{(1)}\ddot{v} - \rho^{(2)}\ddot{\phi} \right] \delta \phi
$$
$$
\left. + \left[\frac{\partial Q_x}{\partial x} + \frac{\partial Q_y}{\partial y} + P_z^{(0)} + f_z^{(0)} - \rho^{(0)}\ddot{w} \right] \delta w \right\} dx\, dy
$$
$$
- \int_{t_0}^{t_1} dt \oint_C \left\{ [N_n - p_n^{(0)}]\delta u_n + [N_{ns} - p_s^{(0)}]\delta u_s \right.
$$
$$
+ [M_n - p_n^{(1)}]\delta \psi_n + [M_{ns} - p_s^{(1)}]\delta \psi_s
$$
$$
\left. + [Q_n - p_z^{(0)}]\delta w \right\} ds = 0, \tag{5.2.2}
$$

where most of the notations are the same as in Eq. (5.1.2). In addition, the transverse shearing forces have now appeared, related to the shearing stresses by

$$[Q_x, Q_y] = \sum_{k=1}^{n} \int_{h_{k-1}}^{h_k} [\sigma_{zx(k)}, \sigma_{zy(k)}]\, dz. \tag{5.2.3}$$

5.2.2 Stress–Strain–Displacement Relations

As in the preceding section, evaluation of the plate stresses $N_x, \ldots, M_x, \ldots$ begins with the set of three-dimensional stress–strain relations in Eqs. (5.1.5) for a material that has a plane of elastic symmetry. The stress–strain relations in Eqs. (5.1.6) for the kth ply and those in Eqs. (5.1.7) for the laminate are also still valid. However, the strain–displacement relations are now different. Corresponding to Eqs. (5.2.1), these now have the form

$$e_{xx} = \frac{\partial u}{\partial x} + z\frac{\partial \psi}{\partial x}$$

$$e_{yy} = \frac{\partial v}{\partial y} + z\frac{\partial \phi}{\partial y}$$

$$e_{xy} = \frac{1}{2}\left(\frac{\partial u}{\partial y} + \frac{\partial v}{\partial x}\right) + \frac{1}{2}z\left(\frac{\partial \psi}{\partial y} + \frac{\partial \phi}{\partial x}\right).$$

The plate strains are thus

$$e_{xx}^{(0)} = \frac{\partial u}{\partial x}$$

$$e_{yy}^{(0)} = \frac{\partial v}{\partial y}$$

$$e_{xy}^{(0)} = \frac{1}{2}\left(\frac{\partial u}{\partial y} + \frac{\partial v}{\partial x}\right)$$

$$e_{xx}^{(1)} = k_x = \frac{\partial \psi}{\partial x} \tag{5.2.4}$$

$$e_{yy}^{(1)} = k_y = \frac{\partial \phi}{\partial y}$$

$$e_{xy}^{(1)} = k_{xy} = \frac{1}{2}\left(\frac{\partial \psi}{\partial y} + \frac{\partial \phi}{\partial x}\right).$$

Equations (5.2.4) are the plate strain–displacement relations. Substitution of these into Eqs. (5.1.7) yields the new plate stress–displacement relations for $N_x, \ldots,$ $M_x, \ldots$.

To derive similar relations for the additional transverse shearing forces Q_x and Q_y, we begin with the same set of three-dimensional stress–strain relations in Eqs. (5.1.5). The pertinent part that should now be singled out consists of the following:

$$\begin{bmatrix} \sigma_{yz} \\ \sigma_{zx} \end{bmatrix} = \begin{bmatrix} c_{44} & c_{45} \\ c_{54} & c_{55} \end{bmatrix} \begin{bmatrix} 2e_{yz} \\ 2e_{zx} \end{bmatrix}. \tag{5.2.5}$$

The linear shearing strains are simply

$$e_{yz} = e_{yz}^{(0)} = \frac{1}{2}\left(\phi + \frac{\partial w}{\partial y}\right)$$

$$e_{zx} = e_{zx}^{(0)} = \frac{1}{2}\left(\psi + \frac{\partial w}{\partial x}\right). \qquad (5.2.6)$$

Applying Eqs. (5.2.5) to the kth ply and substituting the result into Eqs. (5.2.3), we find the following plate stress-strain relations:

$$\begin{bmatrix} Q_y \\ Q_x \end{bmatrix} = \kappa \begin{bmatrix} A_{44} & A_{45} \\ A_{45} & A_{55} \end{bmatrix} \begin{bmatrix} 2e_{yz}^{(0)} \\ 2e_{zx}^{(0)} \end{bmatrix}, \qquad (5.2.7)$$

where $A_{44}, \dots$ are elements of the same A_{ij} as given in the last section, and κ is the shear factor. It was shown in our early work (Yu 1959a,b) that a value of $\kappa = 1$ is a good approximation for the vibration analysis of ordinary sandwich plates. We propose to adopt the same value of $\kappa = 1$ also for the vibration analysis of laminated composite plates. Substitution of Eqs. (5.2.6) into (5.2.7) yields the plate stress–displacement relations for Q_x and Q_y. By means of these and similar relations for $N_x, \dots, M_x, \dots$, the stress equations of motion may be transformed readily into the displacement equations of motion.

5.2.3 Symmetric Laminate

Since $\rho^{(1)} = 0$ and $B_{ij} = 0$ for a symmetric laminate, extension and flexure also become uncoupled in the refined equations. In fact, the displacement equations for extension remain the same as Eqs. (5.1.10) in the classical case. The transverse shear effect affects flexure only. The displacement equations for flexure are now

$$D_{11}\frac{\partial^2 \psi}{\partial x^2} + 2D_{16}\frac{\partial^2 \psi}{\partial x \partial y} + D_{66}\frac{\partial^2 \psi}{\partial y^2} + D_{16}\frac{\partial^2 \phi}{\partial x^2} + (D_{12} + D_{66})\frac{\partial^2 \phi}{\partial x \partial y} + D_{26}\frac{\partial^2 \phi}{\partial y^2}$$

$$- \kappa\left[A_{55}\left(\psi + \frac{\partial w}{\partial x}\right) + A_{45}\left(\phi + \frac{\partial w}{\partial y}\right)\right] + P_x^{(1)} + f_x^{(1)} - \rho^{(2)}\ddot{\psi} = 0$$

$$D_{16}\frac{\partial^2 \psi}{\partial x^2} + (D_{12} + D_{66})\frac{\partial^2 \psi}{\partial x \partial y} + D_{26}\frac{\partial^2 \psi}{\partial y^2} + D_{66}\frac{\partial^2 \phi}{\partial x^2} + 2D_{26}\frac{\partial^2 \phi}{\partial x \partial y} + D_{22}\frac{\partial^2 \phi}{\partial y^2}$$

$$- \kappa\left[A_{45}\left(\psi + \frac{\partial w}{\partial x}\right) + A_{44}\left(\phi + \frac{\partial w}{\partial y}\right)\right] + P_y^{(1)} + f_y^{(1)} - \rho^{(2)}\ddot{\phi} = 0$$

$$\kappa\left[A_{55}\left(\frac{\partial \psi}{\partial x} + \frac{\partial^2 w}{\partial x^2}\right) + A_{45}\left(\frac{\partial \psi}{\partial y} + \frac{\partial \phi}{\partial x} + 2\frac{\partial^2 w}{\partial x \partial y}\right) + A_{44}\left(\frac{\partial \phi}{\partial y} + \frac{\partial^2 w}{\partial y^2}\right)\right]$$

$$+ P_z^{(0)} + f_z^{(0)} - \rho^{(0)}\ddot{w} = 0. \qquad (5.2.8)$$

These have essentially the same form as Mindlin's equations (1951b) for a crystal plate, including his equations (1951a) for an isotropic, homogeneous plate as a special case. We note again as in the preceding section that flexural motion can be excited by surface tractions and body forces in the x- and y-directions, as well as in the z-direction.

5.3   Flexural Vibration of a Symmetric Laminate: Useful Ranges of Equations

For a laminated composite plate, which is anisotropic, it is important to recognize the greater variety of modes of free vibration that are possible than in an isotropic plate. Our coverage will be restricted to a symmetric laminate for which flexure becomes uncoupled from extension. The ranges of applicability of the classical and refined equations for flexure will be determined (Yu 1992b).

5.3.1   Applications of Classical Equation

For an orthotropic symmetric laminate in the absence of surface and body forces, the classical Eq. (5.1.9) reduces to

$$
D_{11}\frac{\partial^4 w}{\partial x^4} + 2(D_{12} + 2D_{66})\frac{\partial^4 w}{\partial x^2 \partial y^2}
$$

$$
+ D_{22}\frac{\partial^4 w}{\partial y^4} + \rho^{(0)}\ddot{w} - \rho^{(2)}\nabla^2\ddot{w} = 0. \tag{5.3.1}
$$

For plane–strain flexural modes in an infinite laminate, we have the following two choices:

$$
w = W \sin \xi x\, e^{i\omega t}, \qquad w = W \sin \eta y\, e^{i\omega t}. \tag{5.3.2}
$$

When these are substituted into Eq. (5.3.1), we find, respectively,

$$
\omega = \xi^2 \left[\frac{D_{11}}{\rho^{(0)} + \rho^{(2)}\xi^2}\right]^{1/2}, \qquad \omega = \eta^2 \left[\frac{D_{22}}{\rho^{(0)} + \rho^{(2)}\eta^2}\right]^{1/2}, \tag{5.3.3}
$$

which are two of the lowest flexural frequencies. Since D_{11} can be, say, ten times greater than D_{22}, the ratio between the two frequencies can be greater than 3 for the same ξ and η. We recognize immediately the possibility of having to deal with all of these modes and frequencies in a dynamic and vibration analysis.

The above results for an infinite symmetric laminate are also readily applicable to a laminate strip that has a finite length ℓ in the x- or y-direction, with simply supported edges at x or $y = 0, \ell$, but extends to infinity in the y- or x-direction. Analysis of two-dimensional flexural modes in laminates of infinite or finite sizes can similarly be carried out by means of Eq. (5.3.1), or by means of the original variational equation of motion. These will be demonstrated in the next chapter.

5.3.2 Applications of Refined Equations

In addition to the lowest flexural branches of the frequency spectrum of an infinite symmetric laminate, the refined flexural Eqs. (5.2.8) also will yield the lowest thickness-shear branches, similar to the situation with an infinite, single-layered isotropic plate. For an orthotropic symmetric laminate, we have $A_{45} = D_{16} = D_{26} = 0$. By further neglecting the surface and body forces, Eqs. (5.2.8) become

$$D_{11}\frac{\partial^2 \psi}{\partial x^2} + D_{66}\frac{\partial^2 \psi}{\partial y^2} + (D_{12} + D_{66})\frac{\partial^2 \phi}{\partial x \partial y} - \kappa A_{55}\left(\psi + \frac{\partial w}{\partial x}\right) - \rho^{(2)}\ddot{\psi} = 0$$

$$(D_{12} + D_{66})\frac{\partial^2 \psi}{\partial x \partial y} + D_{66}\frac{\partial^2 \phi}{\partial x^2} + D_{22}\frac{\partial^2 \phi}{\partial y^2} - \kappa A_{44}\left(\phi + \frac{\partial w}{\partial y}\right) - \rho^{(2)}\ddot{\phi} = 0$$

$$\kappa\left[A_{55}\left(\frac{\partial \psi}{\partial x} + \frac{\partial^2 w}{\partial x^2}\right) + A_{44}\left(\frac{\partial \phi}{\partial y} + \frac{\partial^2 w}{\partial y^2}\right)\right] - \rho^{(0)}\ddot{w} = 0.$$

$$(5.3.4)$$

These now have exactly the same form as Mindlin's equations (1951b) for a crystal plate.

For plane–strain modes in an infinite laminate, we may take the displacements in either of the following two combinations:

$$w = W \sin \xi x\, e^{i\omega t}, \qquad \phi = 0, \qquad \psi = \Psi \cos \xi x\, e^{i\omega t}$$

$$w = W \sin \eta y\, e_{i\omega t}, \qquad \psi = 0, \qquad \phi = \Phi \cos \eta y\, e^{i\omega t}.$$

Substitution of these in Eqs. (5.3.4) yields, respectively,

$$\frac{\rho^{(2)}}{\kappa A_{55}}\omega^4 - \left[1 + \left(\frac{D_{11}}{\kappa A_{55}} + \frac{\rho^{(2)}}{\rho^{(0)}}\right)\xi^2\right]\omega^2 + \frac{D_{11}}{\rho^{(0)}}\xi^4 = 0 \qquad (5.3.5)$$

$$\frac{\rho^{(2)}}{\kappa A_{44}}\omega^4 - \left[1 + \left(\frac{D_{22}}{\kappa A_{44}} + \frac{\rho^{(2)}}{\rho^{(0)}}\right)\eta^2\right]\omega^2 + \frac{D_{22}}{\rho^{(0)}}\eta^4 = 0, \qquad (5.3.6)$$

from which the frequencies are given by

$$\omega_{1,2}^2 = \frac{1}{2\rho^{(0)}\rho^{(2)}}[D_{11}\rho^{(0)}\xi^2 + \kappa A_{55}(\rho^{(0)} + \rho^{(2)}\xi^2) \pm (\{D_{11}\rho^{(0)}\xi^2$$

$$+ \kappa A_{55}(\rho^{(0)} + \rho^{(2)}\xi^2)\}^2 - 4\kappa A_{55}D_{11}\rho^{(0)}\rho^{(2)}\xi^4)^{\frac{1}{2}}] \qquad (5.3.7)$$

$$\omega_{3,4}^2 = \text{same as above with } D_{11} \text{ replaced by } D_{22}, \ A_{55} \text{ by } A_{44}, \text{ and } \xi \text{ by } \eta. \quad (5.3.8)$$

For a given ξ or η, each of Eqs. (5.3.7) and (5.3.8) yields a lower frequency that corresponds to a predominantly flexural mode, and a higher frequency that corresponds to a predominantly thickness-shear mode. Thus, more modes and frequencies of flexural vibrations are generated by the refined equations than by the classical equations, just as in the case of an isotropic plate.

The cutoff frequencies are obtained from Eqs. (5.3.5) and (5.3.6) by setting ξ and η equal to 0. The results are, respectively,

$$\omega = 0, \quad \left(\frac{\kappa A_{55}}{\rho^{(2)}}\right)^{1/2}; \qquad \omega = 0, \quad \left(\frac{\kappa A_{44}}{\rho^{(2)}}\right)^{1/2}.$$

The vanishing frequencies correspond to a transverse translation; the other two are simple thickness-shear frequencies. In general, a thickness-shear branch becomes more important as its cutoff frequency gets to be lower. Moreover, while the effect of rotatory inertia may be small on the lowest flexural branches themselves, it must be retained if the thickness-shear branches are to be included, for mathematical reasons. We have made similar and other observations elsewhere before (Yu 1992a, 1992b).

5.3.3   Effects of Transverse Shear and Rotatory Inertia

As in the cases of homogeneous and sandwich plates, Eqs. (5.3.5) and (5.3.6) have been arranged to appear in such a form that the effects of transverse shear and rotatory inertia can be suppressed readily by letting $\kappa = \infty$ and $\rho^{(2)} = 0$, respectively. Suppression of the transverse shear effect reduces these equations to Eqs. (5.3.3), which are the results of the classical equations. By comparing the results of the classical and refined equations, it can be shown readily that the transverse shear effect becomes negligible under the conditions

$$\frac{\varrho^{(2)}}{\kappa A_{55}}\omega^2 \ll 1, \qquad \frac{D_{11}}{\kappa A_{55}}\xi^2 \ll 1;$$

$$\frac{\varrho^{(2)}}{\kappa A_{44}}\omega^2 \ll 1, \qquad \frac{D_{22}}{\kappa A_{44}}\eta^2 \ll 1. \tag{5.3.9}$$

Similarly, the rotatory inertia effect can be shown to be negligible under the conditions

$$\frac{\varrho^{(2)}}{\varrho^{(0)}}\xi^2 \ll 1, \qquad \frac{\varrho^{(2)}}{\varrho^{(0)}}\eta^2 \ll 1. \tag{5.3.10}$$

The refined flexural equations also may be applied to the analysis of still other types of flexural modes of vibration in infinite and finite laminates.

5.4   Extensional Vibration of a Symmetric Laminate: Useful Ranges of Equations

Extensional vibration of a laminated composite plate has not received any attention in the literature until very recently (Yu 1992b), although that of a single-layered isotropic plate of the classical type was treated by Love (1927) many years ago. We start with Eqs. (5.1.10), the displacement equations of motion for extension of a symmetric laminate. To further simplify, let us consider an orthotropic laminate for which $A_{16} = A_{26} = 0$. In the absence of surface tractions and body forces, Eqs. (5.1.10) finally reduce to

$$A_{11}\frac{\partial^2 u}{\partial x^2} + A_{66}\frac{\partial^2 u}{\partial y^2} + (A_{12} + A_{66})\frac{\partial^2 v}{\partial x \partial y} - \rho^{(0)}\ddot{u} = 0$$

$$(A_{12} + A_{66})\frac{\partial^2 u}{\partial x \partial y} + A_{66}\frac{\partial^2 v}{\partial x^2} + A_{22}\frac{\partial^2 v}{\partial y^2} - \rho^{(0)}\ddot{v} = 0. \tag{5.4.1}$$

As in the flexural case, there can be a greater variety of extensional modes in an anisotropic laminate than in an isotropic plate. For instance, for one-dimensional extensional modes in an infinite orthotropic symmetric laminate, we can have the choice of the following combinations:

$$u = U \sin \xi x\, e^{i\omega t}, \qquad v = 0$$

$$u = U \sin \eta y\, e^{i\omega t}, \qquad v = 0$$

$$u = 0, \qquad v = V \sin \xi x\, e^{i\omega t}$$

$$u = 0, \qquad v = V \sin \eta y\, e^{i\omega t}, \tag{5.4.2}$$

where the wave number ξ or η is equal to $2\pi/L$, L being the wavelength in the x- or y-direction. Substitution of Eqs. (5.4.2) into (5.4.1) yields the following frequencies, respectively:

$$\omega = \xi \left[\frac{A_{11}}{\rho^{(0)}}\right]^{1/2}$$

$$\omega = \eta \left[\frac{A_{66}}{\rho^{(0)}}\right]^{1/2}$$

$$\omega = \xi \left[\frac{A_{66}}{\rho^{(0)}}\right]^{1/2}$$

$$\omega = \eta \left[\frac{A_{22}}{\rho^{(0)}}\right]^{1/2}.$$

All of these are proportional to the wave number as in a single-layered isotropic plate. But they now represent at least three different frequency values for a laminate instead of just one for an isotropic plate. As an example, consider the T300/5208 cross-ply laminates described by Tsai and Hahn (1980). Since they have stiffnesses in the ratio of $A_{11} : A_{22} : A_{66} = 162.75 : 29.39 : 7.70$, the corresponding frequencies bear the ratios $4.60 : 1.96 : 1$, thus covering a wide frequency range. In the dynamic and vibration analysis of such a laminate, we realize immediately the importance of recognizing the possibility of having to deal with all these modes and frequencies, as in the flexural case.

The first and last of the above four sets of results for an infinite laminate are also readily applicable to a laminate strip that has a finite length ℓ in the x- or y-direction, with fixed edges at x or $y = 0, \ell$, but extends to infinity in the y- or

x-direction; it is only necessary to let ξ or η equal to $n\pi/\ell$, where n is the number of half-waves contained in the length ℓ of the strip.

As an example in two dimensions we take

$$u = U \, \sin \, \xi x \, \sin \, \eta y \, e^{i\omega t}, \qquad v = V \, \cos \, \xi x \, \cos \, \eta y \, e^{i\omega t}.$$

Substitution of these into Eqs. (5.4.1) yields a frequency equation from which two different frequencies are obtained for any given pair of values of ξ and η. The results are applicable to both an infinite laminate and a rectangular laminate with fixed edges. For the infinite laminate, we let $\xi = 2\pi/a$ and $\eta = 2\pi/b$, where a and b are the wavelengths in the x- and y-directions, respectively. For a rectangular laminate, we replace these by $\xi = m\pi/a$ and $\eta = n\pi/b$, where a and b are now the lengths of the plate in the x- and y-directions, and m and n are the number of half-waves contained in a and b, respectively.

References

Agarwal, B.D. and L.J. Broutman. (1990) *Analysis and Performance of Fiber Composites*. John Wiley, New York.

Dudek, T.J. (1970) Young's and Shear Moduli of Unidirectional Composites by a Resonant Beam Method. *Journal of Composite Materials*, Vol. 4, pp. 232–241.

Hearmon, R.F.S. (1961) *An Introduction to Applied Anisotropic Elastic Body*. Oxford University Press, Oxford, England.

Jones, R.M. (1975) *Mechanics of Composite Materials*. McGraw-Hill, New York.

Lekhnitskii, S.G. (1950) *Theory of Elasticity of an Anisotropic Elastic Body*. Government Publishing House for Technical-Theoretical Works, Moscow and Leningrad, Russia. (English translation, Holden-Day, 1963.)

Love, A.E.H. (1927) *The Mathematical Theory of Elasticity*, 4th Ed. Cambridge University Press, Cambridge, England. Also, Dover Publications, New York, 1944.

Mindlin, R.D. (1951a) Influence of Rotatory Inertia and Shear on Flexural Motions of Isotropic, Elastic Plates. *Journal of Applied Mechanics*, Vol. 18, pp. 31–38.

Mindlin, R.D. (1951b) Thickness-Shear and Flexural Vibrations of Crystal Plates. *Journal of Applied Physics*, Vol. 22, pp. 316–323.

Tsai, S.W. and H.T. Hahn. (1980) *Introduction to Composite Materials*. Technomic Publishing Co., Lancaster, Pennsylvania.

Vinson, J.R. and R.L. Sierakowski. (1987) *The Behavior of Structures Composed of Composite Materials*. Martinus Nijhoff Publishers, Boston, Massachusetts.

Whitney, J.M. (1987) *Structural Analysis of Laminated Anisotropic Plates*. Technomic Publishing Co., Lancaster, Pennsylvania.

Whitney, J.M. and N.J. Pagano. (1970) Shear Deformation in Heterogeneous Anisotropic Plates. *Journal of Applied Mechanics*, Vol. 37, pp. 1031–1036.

Yu, Y.Y. (1959a) A New Theory of Elastic Sandwich Plates—One-Dimensional Case. *Journal of Applied Mechanics*, Vol. 26, pp. 415–421.

Yu, Y.Y. (1959b) Simple Thickness-Shear Modes of Vibration of Infinite Sandwich Plates. *Journal of Applied Mechanics*, Vol. 26, pp. 679–681.

Yu, Y.Y. (1989) Dynamics and Vibration of Layered Plates and Shells—A Perspective from Sandwiches to Laminated Composites. Bound volume for the 30th AIAA/ASME/ASCE/AHS/ASC Structures, Structural Dynamics, and Materials Conference, pp. 2236–2245.

Yu, Y.Y. (1992a) Dynamics and Vibration of Layered Plates and Shells–A Perspective from Sandwiches to Laminated Composites (Russian translation), *Mechanics of Composite Materials, Vol. 2, Construction of Composites, Proceedings of the First Soviet-American Symposium*, pp. 99–108. Riga, Latvia.

Yu, Y.Y. (1992b) Extensional and Flexural Vibrations of Laminated Composite Plates. In: *Proceedings of the 7th Technical Conference on Composite Materials*, American Society for Composites, pp. 962–971. Technomic Publishing Co., Lancaster, Pennsylvania.

Yu, Y.Y. (1995) Some Recent Advances in Linear and Nonlinear Dynamical Modeling of Elastic and Piezoelectric Plates. *Journal of Intelligent Material Systems and Structures*, Vol. 6, pp. 237–254.

6

Linear Vibrations Based on Plate Equations

In Chapters 3 through 5, linear dynamical modeling has been developed for homogeneous, sandwich, and laminated composite plates. Now that plate equations are available, they can be applied directly to the linear vibration analysis of a finite-sized as well as an infinite plate. In this chapter, some topics on free and forced linear vibration analysis of finite plates have been selected.

In Section 6.1, free flexural vibrations of simply supported plates are discussed. This is to call attention to the important and useful fact that results for free vibrations of an infinite plate often can be applied to a simply supported plate. In Section 6.2, the analysis is next extended to free flexural vibrations of clamped plates. The discussion also serves to demonstrate the proper use of the variational equation of motion in an approximate vibration analysis (Yu and Lai 1967). Finally, in Section 6.3 we treat forced flexural vibration of homogeneous and sandwich plates in plane strain that involve time-dependent boundary conditions (Yu 1960).

Without going into any details, we mention here some of our early discussions of other related topics on linear vibrations. These include two-layered plates (Ren and Yu 1967) and the damping of two-layered as well as sandwich plates (Yu 1962; Yu and Ren 1965).

6.1 Free Flexural Vibration of Plates with Simply Supported Edges

In the linear dynamical modeling of various types of plates, free vibration of an infinite plate often is studied for the purpose of determining the ranges of applicability of the plate equations that are developed. Such vibration studies of an infinite plate serve still another very useful purpose, namely, the results obtained are usually also applicable to finite-sized plates with simply supported edges. This is demonstrated here for homogeneous and sandwich plates, although some related examples of laminated composite plates have already been given in the preceding chapter.

6.1.1   Homogeneous Plate in Plane Strain

Consider the use of Mindlin's refined equations for flexure of a homogeneous plate that include the transverse shear effect. For a plate in plane strain, these reduce to Eqs. (3.5.4), which involve only x-dependence and become analogous to the Timoshenko beam equations (3.2.11). For free vibration of an infinite plate, the plate displacements were taken in the form

$$w = W \sin \xi x \, e^{i\omega t}, \quad \psi = \Psi \cos \xi x \, e^{i\omega t}, \tag{6.1.1}$$

where $\xi = 2\pi/L$ is the wave number and L is the wavelength. These expressions are immediately applicable to a finite plate in plane strain with simply supported edges at $x = 0$ and $x = \ell$; it is only necessary to let $\xi = n\pi/\ell$, where n is the number of half-waves in the plate length ℓ. The above expressions then are easily seen to satisfy the following boundary conditons for the simply supported edges:

$$w = 0 \quad \text{and} \quad M_x = D\frac{\partial \psi}{\partial x} = 0 \quad (\text{for } x = 0 \text{ and } x = \ell).$$

With the new interpretation of ξ, therefore, the frequency equation for an infinite plate also becomes applicable to a simply supported plate.

The free vibration analysis of a sandwich plate in plane strain may be carried out similarly; but, instead, we shall next demonstrate with the two-dimensional case of a rectangular sandwich plate.

6.1.2   Rectangular Sandwich Plate

Toward the end of Section 3.5, we discussed briefly the free vibration of an infinite homogeneous plate with phase reversals in both x- and y-directions. As was pointed out, plate displacements are taken in the typical form

$$w = W \sin \xi_x x \sin \xi_y y \, e^{i\omega t} \quad \text{with} \quad \xi_x = \frac{2\pi}{a} \quad \text{and} \quad \xi_y = \frac{2\pi}{b},$$

where a and b are the wavelengths in the x- and y-directions, respectively. This also may be taken in the slightly different form

$$w = W \sin \lambda_x \frac{x}{h} \sin \lambda_y \frac{y}{h} \, e^{i\omega t} \quad \text{with} \quad \lambda_x = \frac{2\pi h}{a} \quad \text{and} \quad \lambda_y = \frac{2\pi h}{b}$$

and applied to an infinite sandwich plate. For instance, based on the general refined equations of a sandwich including transverse shear in all layers derived in Section 4.1, plate displacements are taken in the form (Yu 1962)

$$w = W \sin \frac{\lambda_x x}{h} \sin \frac{\lambda_y y}{h} e^{i\omega t}$$

$$\psi_1 = \Psi_1 \cos \frac{\lambda_x x}{h} \cos \frac{\lambda_y y}{h} e^{i\omega t}$$

$$\psi_2 = \Psi_2 \cos \frac{\lambda_x x}{h} \cos \frac{\lambda_y y}{h} e^{i\omega t} \qquad (6.1.2)$$

$$\phi_1 = \Phi_1 \sin \frac{\lambda_x x}{h} \sin \frac{\lambda_y y}{h} e^{i\omega t}$$

$$\phi_2 = \Phi_2 \sin \frac{\lambda_x x}{h} \sin \frac{\lambda_y y}{h} e^{i\omega t}.$$

Two frequency equations then are obtained for an isotropic sandwich plate: one of them yields three frequencies and the other yields two more.

We recall that, according to Eqs. (4.1.5), for a rectangular sandwich plate with simply supported edges at $x = 0, a$ and $y = 0, b$, the boundary conditions are taken as

$$w = \phi_1 = \phi_2 = M_{x1} + h_1(N_{x3} - N_{x2})$$
$$= M_{x2} + M_{x3} - h_1(N_{x3} - N_{x2}) = 0 \qquad (x = 0, a)$$
$$w = \psi_1 = \psi_2 = M_{y1} + h_1(N_{y3} - N_{y2})$$
$$= M_{y2} + M_{y3} - h_1(N_{y3} - N_{y2}) = 0 \qquad (y = 0, b).$$

It is easily verified that all of these conditions are satisfied by adopting the same plate displacements in Eqs. (6.1.2) and by now taking

$$\lambda_x = \frac{m\pi h}{a}, \qquad \lambda_y = \frac{n\pi h}{b} \qquad (m, n, = 1, 2, 3, \ldots),$$

where m and n designate the number of half-waves contained in the dimensions a and b, respectively, of the rectangular plate. With these new definitions of λ_x and λ_y, results for an infinite sandwich plate become immediately applicable to a simply supported rectangular sandwich plate.

6.2   Free Flexural Vibration of Plates with Clamped Edges

The free flexural vibration of a clamped sandwich plate in plane strain will be analyzed on the basis of the simplified refined equations of a sandwich with membrane facings developed in Section 4.2. The sandwich plate has a span in the x-direction, with clamped edges at $x = 0, \ell$, and it includes the homogeneous plate as a special case (Yu and Lai 1967).

6.2.1   Variational Equation of Motion for Plate in Plane Strain

We start with the variational equation of motion in Eq. (4.2.3), which reduces to, in the case of plane strain,

$$\int_{t_0}^{t_1} dt \int_0^{\ell} \left(\left\{ \frac{\partial}{\partial x}[M_{x1} + h_1(N_{x3} - N_{x2})] - \cdots \right\} \delta\psi + \{\cdots\} \delta w \right) dx \, dt$$

$$- \int_{t_0}^{t_1} \{\cdots\} \, dt = 0. \tag{6.2.1}$$

The second intergral in this equation is associated with the boundary conditions. It vanishes when the boundary conditions for the clamped edges are chosen to be

$$w = 0, \qquad \psi = 0 \qquad (x = 0, \ell). \tag{6.2.2}$$

By writing the plate stresses in terms of the plate displacements, and by ignoring the external force terms as well as the integration with respect to time, Eq. (6.2.1) reduces to

$$\int_0^{\ell} \left\{ \left[\left(\frac{2}{3} \epsilon_1 h_1^3 + 2\epsilon_2 h_2 h_1^2 \right) \frac{\partial^2 \psi}{\partial x^2} - 2\kappa_1 G_1 h_1 \left(\psi + \frac{\partial w}{\partial x} \right) \right. \right.$$

$$\left. - \left(\frac{2}{3} \rho_1 h_1^3 + 2\rho_2 h_2 h_1^2 \right) \ddot{\psi} \right] \delta\psi$$

$$\left. + \left[2\kappa_1 G_1 h_1 \left(\frac{\partial \psi}{\partial x} + \frac{\partial^2 w}{\partial x^2} \right) - 2(\rho_1 h_1 + \rho_2 h_2)\ddot{w} \right] \delta w \right\} dx = 0. \tag{6.2.3}$$

For free vibration, we let

$$w = W(x) \, h \, e^{i\omega t}, \qquad \psi = \Psi(x) \, e^{i\omega t},$$

and Eq. (6.2.3) finally becomes

$$\int_0^{\ell} \{ [\Omega\Psi(1 + 3r_\rho r_h) + (r_1 + 3r_2 r_h)\Psi'' - 3\kappa_1(1 + r_h)^2(\Psi + W')] \delta\Psi$$

$$+ 3(1 + r_h)^2[\Omega W(1 + r_\rho r_h) + \kappa_1(\Psi' + W'')] \delta W \} \, dx = 0. \tag{6.2.4}$$

6.2.2   Plate and Beam Equations

The equations of motion associated with Eq. (6.2.4) may be written by setting the coefficients of $\delta\Psi$ and δW equal to zero. Thus,

$$\Omega\Psi(1 + 3r_\rho r_h) + (r_1 + 3r_2 h_2)\Psi'' - 3\kappa_1(1 + r_h)^2(\Psi + W') = 0$$

$$\Omega W(1 + r_\rho r_h) + \kappa_1(\Psi' + W'') = 0, \tag{6.2.5}$$

which also can be written directly from Eqs. (4.2.9).

Often, either W or Ψ is eliminated, resulting in a single uncoupled equation of the form

$$\left[\frac{1}{\kappa_1}\,\Omega^2(1+3r_\rho r_h)(1+r_\rho r_h) - 3\Omega(1+r_h)^2(1+r_\rho r_h)\right]W$$

$$+\left[\frac{1}{\kappa_1}\Omega(1+r_\rho r_h)(r_1+3r_2 r_h) + \Omega(1+3r_\rho r_h)\right]W''$$

$$+\ (r_1+3r_2 r_h)W'''' = 0, \qquad\qquad (6.2.6)$$

which, for instance, no longer includes Ψ. When the transverse shear effect is suppressed by setting κ_1 equal to infinity and the effect of rotatory inertia is neglected by replacing $(1+3r_\rho r_h)$ by zero, Eq. (6.2.6) becomes

$$3\Omega(1+r_h)^2(1+r_\rho r_h)W - (r_1+3r_2 r_h)W'''' = 0. \qquad (6.2.7)$$

This equation is then of the classical type.

Finally, by setting $r_h = 0$, Eqs. (6.2.4) through (6.2.7) become equations for a single-layered homogeneous plate. Equation (6.2.6) is essentially reducible to the usual Timoshenko beam equation. Apparently, the latter name is somewhat a misnomer because the equation by itself is incomplete. Similarly, Eq. (6.2.7) is reducible to the classical beam equation.

6.2.3 Free Vibration of a Plate with Clamped Edges: Exact Solution

The ordinary differential Eqs. (6.2.5), or Eq. (6.2.6) together with either one of Eqs. (6.2.5), may be solved exactly in the usual manner (Yu and Lai 1967). The general solution has the form

$$W = C_1 \cosh\frac{\alpha x}{h} + C_2 \sinh\frac{\alpha x}{h} + C_3 \cos\frac{\beta x}{h} + C_4 \sin\frac{\beta x}{h}$$

$$\Psi = R_1\left(C_1 \sinh\frac{\alpha x}{h} + C_2 \cosh\frac{\alpha x}{h}\right) + R_2\left(-C_3 \sin\frac{\beta x}{h} + C_4 \cos\frac{\beta x}{h}\right),$$

where

$$\alpha^2 = \frac{-B + (B^2 - 4AC)^{1/2}}{2A}$$

$$\beta^2 = \frac{+B + (B^2 - 4AC)^{1/2}}{2A}$$

$$A = r_1 + 3r_2 r_h$$

$$B = \Omega\left[(1+3r_\rho r_h) + \frac{1}{\kappa_1}(1+r_\rho r_h)(r_1+3r_2 r_h)\right]$$

$$C = \frac{1}{\kappa_1}\Omega^2(1 + 3r_\rho r_h)(1 + r_\rho r_h) - 3\Omega(1 + r_h)^2(1 + r_\rho r_h)$$

$$R_1 = \frac{1}{\alpha}\left[-\alpha^2 - \frac{\Omega}{\kappa_1}(1 + r_\rho r_h)\right]$$

$$R_2 = \frac{1}{\beta}\left[-\beta^2 + \frac{\Omega}{\kappa_1}(1 + r_\rho r_h)\right].$$

For a sandwich plate with both edges clamped, for which $w = \psi = 0$ at $x = 0, \ell$, we obtain the following frequency equation:

$$(R_1^2 - R_2^2)\,\sinh\frac{\alpha\ell}{h}\,\sin\frac{\beta\ell}{h} + 2R_1 R_2\left(1 - \cosh\frac{\alpha\ell}{h}\,\cos\frac{\beta\ell}{h}\right) = 0. \quad (6.2.8)$$

When the effects of transverse shear and rotatory inertia are suppressed, this reduces to, for the classical case,

$$1 - \cosh\left[\frac{3\Omega(1 + r_h)^2(1 + r_\rho r_h)}{r_1 + 3r_2 r_h}\right]^{1/4}\frac{\ell}{h}$$

$$\times \cos\left[\frac{3\Omega(1 + r_h)^2(1 + r_\rho r_h)}{r_1 + 3r_2 r_h}\right]^{1/4}\frac{\ell}{h} = 0. \quad (6.2.9)$$

With $r_h = 0$, Eq. (6.2.9) becomes the frequency equation for a single-layered classical plate or beam such as may be found in Timoshenko's book (1955).

6.2.4   Free Vibration of a Plate with Clamped Edges: Approximate Solution

The equations of motion also may be solved by means of the approximate method of Galerkin. The variational equation of motion, Eq. (6.2.4), is a good starting point for carrying out such an approximation. The two parts associated with $\delta\Psi$ and δW in the equation are written separately as follows:

$$\int_0^\ell [\Omega\Psi(1 + 3r_\rho r_h) + (r_1 + 3r_2 r_h)\Psi'' - 3\kappa_1(1 + r_h)^2(\Psi + W')]\,\delta\Psi\,dx = 0$$
$$(6.2.10)$$
$$\int_o^\ell [\Omega W(1 + r_\rho r_h) + \kappa_1(\Psi' + W'')]\,\delta W\,dx = 0,$$

which are ready for application of the Galerkin procedure. In contrast to this formulation, it is a common practice first to uncouple the differential equations and then to apply the Galerkin procedure to the single uncoupled equation. In the present problem, Eq. (6.2.6) is the uncoupled equation, and the variational formulation would take the following form:

$$\int_0^\ell \left\{\left[\frac{1}{\kappa_1}\Omega^2(1 + 3r_\rho r_h)(1 + r_\rho r_h) - 3\Omega(1 + r_h)^2(1 + r_\rho r_h)\right]W\right.$$

$$+ \left[\frac{1}{\kappa_1}\Omega(1 + r_\rho r_h)(r_1 + 3r_2 r_h) + \Omega(1 + 3r_\rho r_h)\right] W''$$

$$\left. + (r_1 + 3r_2 r_h)W''''\right\} \, \delta W \, dx = 0, \tag{6.2.11}$$

which, however, is neither proper nor correct from the standpoint of the variational principle, as will be elaborated presently.

Only the fundamental mode will be dealt with for the clamped sandwich plate. The displacements are assumed to be in the form

$$W = A\left(\frac{x}{\ell}\right)^2 \left(1 - \frac{x}{\ell}\right)^2$$

$$\Psi = B\left(\frac{x}{\ell}\right)\left(1 - \frac{x}{\ell}\right)\left(1 - \frac{2x}{\ell}\right), \tag{6.2.12}$$

which satisfy the boundary conditions in Eqs. (6.2.2) and therefore may be used in the Galerkin method. Indeed, the second of Eqs. (6.2.12) further assures that Ψ vanishes at the mid-span $x = \ell/2$, a condition also required of the fundamental mode. When Eqs. (6.2.12) are substituted into Eqs. (6.2.10) and integration is carried out with respect to x, the following frequency equation for the clamped plate is obtained:

$$\left[\frac{1}{\kappa}\Omega^2(1 + 3r_\rho r_h)(1 + r_\rho r_h) - 3\Omega(1 + r_h)^2(1 + r_\rho r_h)\right]\left(\frac{\ell}{h}\right)^4$$

$$- 12\left[\frac{1}{\kappa_1}\frac{7}{2}\Omega(1 + r_\rho r_h)(r_1 + 3r_2 r_h) + \Omega(1 + 3r_\rho r_h)\right]\left(\frac{\ell}{h}\right)^2$$

$$+ 504(r_1 + 3r_2 r_h) = 0. \tag{6.2.13}$$

If the Galerkin approximation is performed on the basis of the uncoupled Eq. (6.2.11), only the assumed W in Eqs. (6.2.12) is needed. When this is substituted into Eq. (6.2.11) and integration is carried out, the frequency equation is found to be

$$\left[\frac{1}{\kappa_1}\Omega^2(1 + 3r_\rho r_h)(1 + r_\rho r_h) - 3\Omega(1 + r_h)^2(1 + r_\rho r_h)\right]\left(\frac{\ell}{h}\right)^4$$

$$- 12\left[\frac{1}{\kappa_1}\Omega(1 + r_\rho r_h)(r_1 + 3r_2 r_h) + \Omega(1 + 3r_\rho r_h)\right]\left(\frac{\ell}{h}\right)^2$$

$$+ 504(r_1 + 3r_2 r_h) = 0. \tag{6.2.14}$$

Although the factor 7/2 constitutes the only difference between Eqs. (6.2.13) and (6.2.14), this will lead to a large discrepancy in the calculated values of the fundamental frequency. The discrepancy is large because the ratio r_2 is usually very large for a sandwich plate.

6.2.5 Numerical Examples

Consider the numerical cases in which we take the ratios $r_1 = 2.2$, $r_2 = 4790$, $r_\rho = 34.4$, $r_\mu = 1683$, $\kappa_1 = 1$, and $r_h = 0$, 0.02, 0.01, and 0.2. The case $r_h = 0$ represents a homogeneous plate, and the other values of r_h are for sandwich plates.

Natural frequencies for these plates with clamped edges were first calculated by solving exactly Eqs. (6.2.8) and (6.2.9), which we recall were derived from the refined and classical theories, respectively. The results are shown in Figures 6.2.1 and 6.2.2. The difference between the two theories diminishes for all homogeneous and sandwich cases as ℓ/h becomes very large. For a homogeneous plate, the difference in general remains small even when ℓ/h is not large, and is indeed insignificant whenever ℓ/h is larger than 25. This is not so for sandwich plates. As is revealed by these figures, the differences between the two theories for sandwich plates can be substantial even for moderate values of ℓ/h. Figure 6.2.2 for $r_h = 0.1$ further shows that the differences also become greater for higher modes. To gain some idea as to the frequency range covered in Figures 6.2.1 and 6.2.2, we remark that for a sandwich plate with aluminum facings and total thickness of 0.3 inch, $\sqrt{\Omega} = 10^{-3}$, 10^{-2}, and 10^{-1} correspond to 18.8, 188, and 1880 hertz, respectively.

The fundamental frequencies of the clamped sandwich plate with $r_h = 0.1$ next were calculated by means of the Galerkin method. Computations were based on Eqs. (6.2.13) and (6.2.14) derived from the proper and improper formulations, respectively. As is shown in Figure 6.2.3, the approximate results thus obtained are higher than the exact ones given by the refined theory. The results of the proper Galerkin approximation yielded by Eq. (6.2.13), however, are rather close to the exact results and differ from the latter by only 8% for $\ell/h = 16$ and 4% for $\ell/h = 100$. On the other hand, as is shown in the same figure, the results of the improper Galerkin approximation given by Eq. (6.2.14) can differ substantially from the exact results of the refined theory, the difference being about 39% for $\ell/h = 100$.

6.3 Forced Flexural Vibration of Homogeneous and Sandwich Plates in Plane Strain

In this section, we treat the general problem of forced flexural vibration of homogeneous and sandwich plates in plane strain (Yu 1960). The analysis is based on the general refined equations of a sandwich plate presented in Section 4.1, which includes the homogeneous plate as a special case. The classical method of separation of variables combined with the Mindlin-Goodman procedure (1950) for treating time-dependent boundary conditions is used. The span of the plate is chosen to be in the x-direction, with edges at $x = 0$ and $x = \ell$.

6.3.1 Basic Equations

In the general refined flexural theory of a sandwich plate that includes transverse shear effect in all layers, the stress equations of motion and boundary conditions

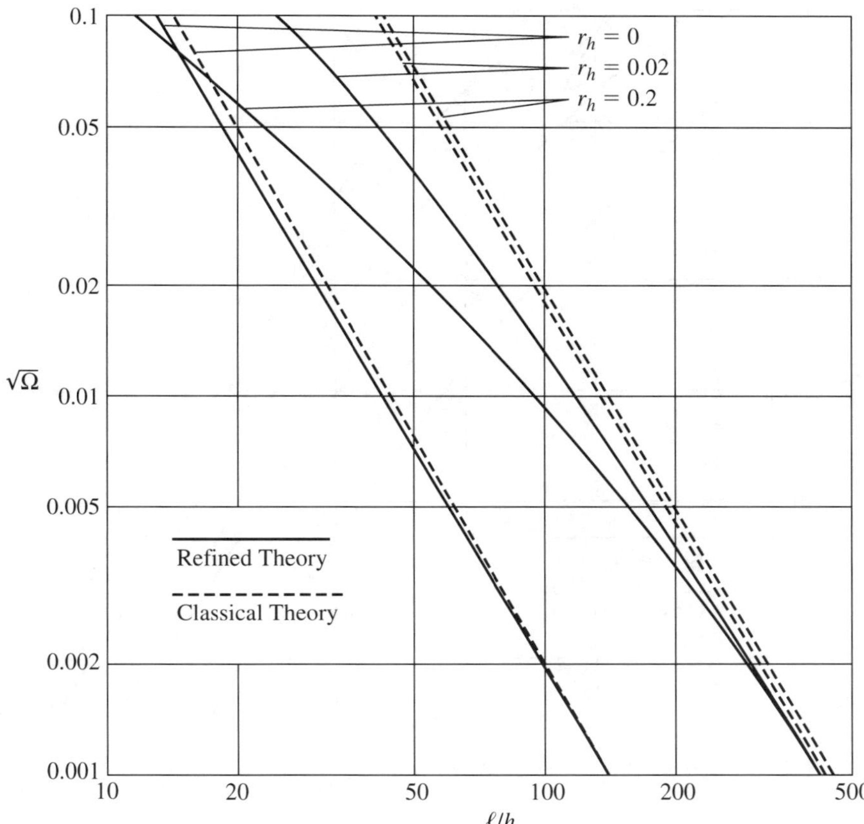

Fig. 6.2.1. Fundamental frequencies of clamped homogeneous ($r_h = 0$) and sandwich ($r_h = 0.02, 0.2$) plates.

are given by Eqs. (4.1.4) and (4.1.5), respectively. A one-dimensional version is obtainable from the first, second, and fifth of these equations. The displacement equations of motion have the following form:

$$(2D_1 + E_2 h_1^2)\psi_1'' - 2G_1(\psi_1 + w') + 6G_2\left(\frac{h_1}{h_2}\right)(\psi_2 + w')$$

$$- h_1(p_x^+ - p_x^-) = \left(\frac{4\rho_1 h_1^3}{3} + \rho_2 h_1^2 h_2\right)\ddot{\psi}_1$$

$$E_2 h_1 h_2 \psi_1'' + \tfrac{2}{3}E_2 h_2^2 \psi_2'' - 2G_2(\psi_2 + w')$$

$$+ h_2(p_x^+ - p_x^-) = \rho_2 h_2^2\left(h_1\ddot{\psi}_1 + \frac{2h_2\ddot{\psi}_2}{3}\right) \tag{6.3.1}$$

$$G_1(\psi_1' + w'') + 2G_2(\psi_2' + w'') + p_z^+ + p_z^- = 2(\rho_1 h_1 + \rho_2 h_2)\ddot{w},$$

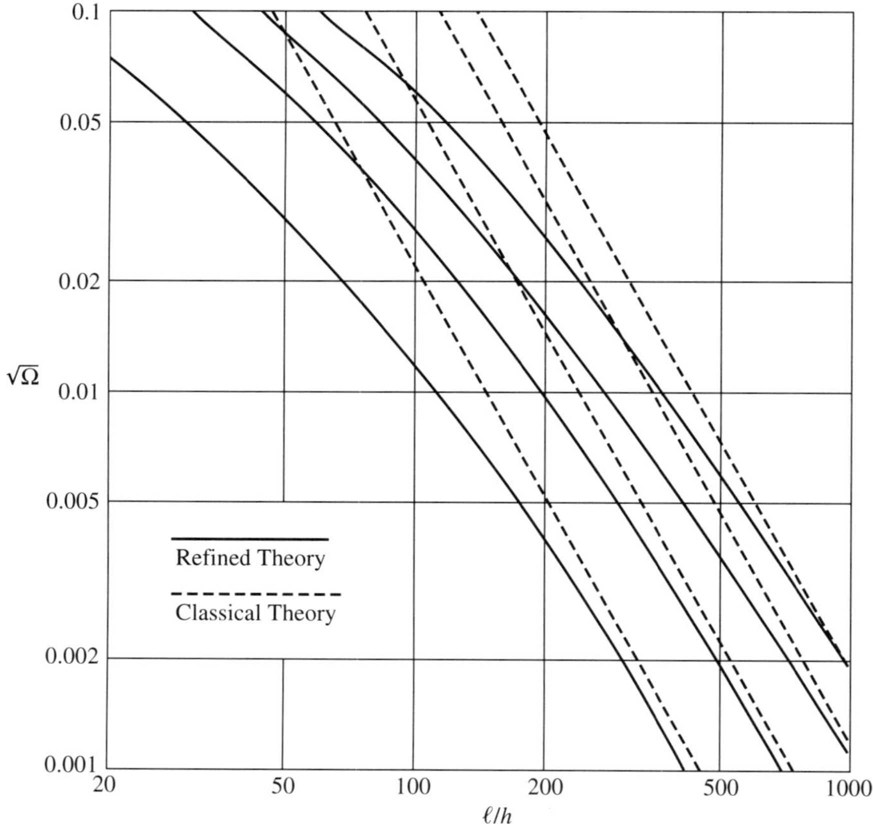

Fig. 6.2.2. Frequencies of first four modes of clamped sandwich plates ($r_h = 0.1$)

in which the body force terms have been neglected. Within the span of the plate, tractions $(p_z^+ + p_z^-)$ and $(p_x^+ - p_x^-)$ at the top and bottom boundary planes are prescribed.

Similarly, the boundary conditions for the edges at $x = 0$ and $x = \ell$ are written from Eqs. (4.1.5):

$$B_1 = M_{x1} + h_1(N_{x3} - N_{x2}) \qquad \text{or} \quad \psi_1 \quad (\text{at } x = 0)$$

$$B_2 = M_{x1} + h_1(N_{x3} - N_{x2}) \qquad \text{or} \quad \psi_1 \quad (\text{at } x = \ell)$$

$$B_3 = M_{x2} + M_{x3} - h_1(N_{x3} - N_{x2}) \ \text{or} \quad \psi_2 \quad (\text{at } x = 0)$$

$$B_4 = M_{x2} + M_{x3} - h_1(N_{x3} - N_{x2}) \ \text{or} \quad \psi_2 \quad (\text{at } x = \ell)$$

$$B_5 = Q_{n1} + Q_{n2} + Q_{n3} \qquad \text{or} \quad w \quad (\text{at } x = 0)$$

$$B_6 = Q_{n1} + Q_{n2} + Q_{n3} \qquad \text{or} \quad w \quad (\text{at } x = \ell).$$

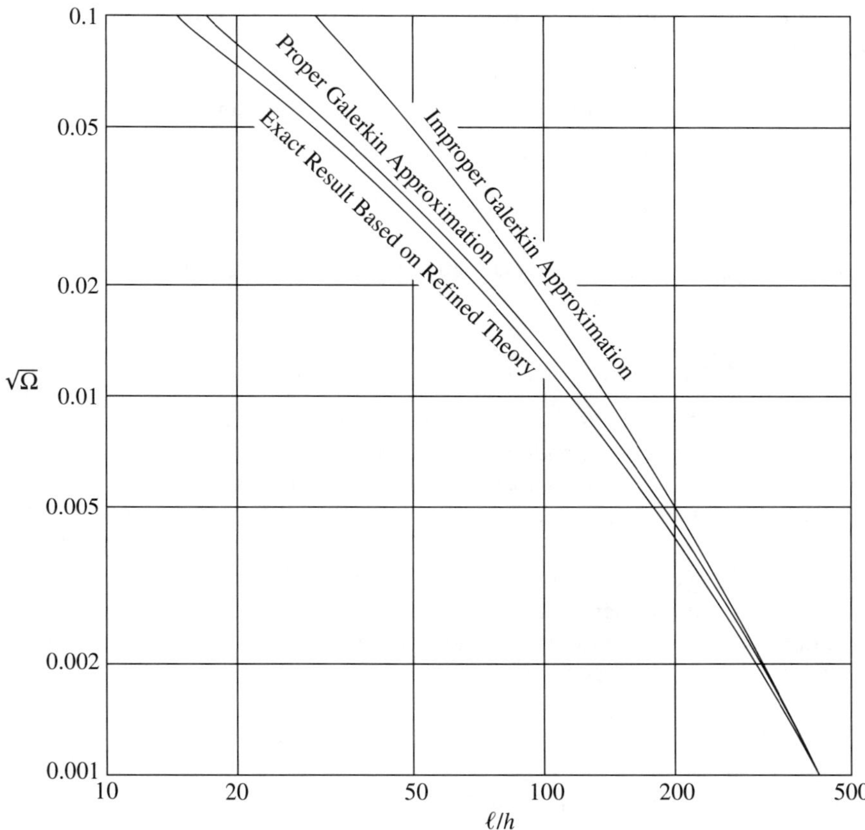

Fig. 6.2.3. Fundamental frequencies of clamped sandwich plates ($r_h = 0.1$): Comparison between exact and approximate results.

Or, in terms of plate displacements,

$$B_1 = (D_1 + 2E_2h_1^2)\psi_1' + E_2h_1h_2\psi_2' \quad \text{or} \quad \psi_1 \qquad (\text{at } x = 0)$$

$$B_2 = (D_1 + 2E_2h_1^2)\psi_1' + E_2h_1h_2\psi_2' \quad \text{or} \quad \psi_1 \qquad (\text{at } x = \ell)$$

$$B_3 = E_2h_1h_2\psi_1' + \tfrac{2}{3}E_2h_2^2\psi_2' \quad \text{or} \quad \psi_2 \qquad (\text{at } x = 0)$$

$$B_4 = E_2h_1h_2\psi_1' + \tfrac{2}{3}E_2h_2^2\psi_2' \quad \text{or} \quad \psi_2 \qquad (\text{at } x = \ell)$$

$$B_5 = G_1(\psi_1 + w') + 2G_2(\psi_2 + w') \quad \text{or} \quad w \qquad (\text{at } x = 0)$$

$$B_6 = G_1(\psi_1 + w') + 2G_2(\psi_2 + w') \quad \text{or} \quad w \qquad (\text{at } x = \ell)$$

Thus $B_1, B_2, \ldots,$ and B_6 always can be written in terms of the plate displacements $\psi_1, \psi_2,$ and w, and the boundary conditions may be written in the general form

$$B_i(\psi_1, \psi_2, w) = f_i(t) \qquad (i = 1, 2, \ldots, 6), \tag{6.3.2}$$

where $f_i(t)$ are prescribed functions of time.

Finally, the initial conditions require the prescription of the following initial displacements and velocities:

$$\psi_1(x, 0) = \psi_{10}(x), \qquad \dot{\psi}_1(x, 0) = \dot{\psi}_{10}(x)$$

$$\psi_2(x, 0) = \psi_{20}(x), \qquad \dot{\psi}_2(x, 0) = \dot{\psi}_{20}(x) \tag{6.3.3}$$

$$w(x, 0) = w_0(x), \qquad \dot{w}(x, 0) = \dot{w}_0(x),$$

where the right-hand side of each condition represents a given function.

6.3.2 Principal Modes and Orthogonality Condition

In the free vibration analysis of a sandwich plate, an infinite set of natural frequencies ω_n and the corresponding principal modes $(\Psi_{1n}, \Psi_{2n}, W_n)$ of the plate, with $n = 1, 2, \ldots$, are determined from the homogeneous equations of motion and homogeneous boundary conditions associated with Eqs. (6.3.1) and (6.3.2). Analysis has been carried out for an infinite sandwich plate in Section 4.4 and extended to the special case of a simply supported sandwich plate in Section 6.1. As mentioned earlier, there is actually a triply infinite set of principal modes in the simply supported case. In the general case, by neglecting the surface traction terms in Eqs. (6.3.1) and by taking

$$\psi_1 = \Psi_1(x)\, e^{i\omega t}, \qquad \psi_2 = \Psi_2(x)\, e^{i\omega t}, \qquad w = W(x)\, e^{i\omega t},$$

the principal modes are found to satisfy the following equations:

$$(2D_1 + E_2 h_1^2)\Psi_{1n}'' - 2G_1(\Psi_{1n} + W_n') + 6G_2\frac{h_1}{h_2}(\Psi_{2n} + W_n')$$

$$= -\omega_n^2\left(\frac{4\rho_1 h_1^3}{3} + \rho_2 h_1^2 h_2\right)\Psi_{1n}$$

$$E_2 h_1 h_2 \Psi_{1n}'' + E_2\frac{2h_2^2}{3}\Psi_{2n}'' - 2G_2(\Psi_{2n} + W_n') \tag{6.3.4}$$

$$= -\omega_n^2 \rho_2 h_2^2\left(h_1\Psi_{1n} + \tfrac{2}{3}h_2\Psi_{2n}\right)$$

$$G_1(\Psi_{1n}' + W_n'') + 2G_2(\Psi_{2n}' + W_n'') = -2\omega_n^2(\rho_1 h_1 + \rho_2 h_2)W_n.$$

The orthogonality condition of the principal modes is very useful in a forced vibration analysis. While this also may be derived from the equations of motion (6.3.4), the derivation is much simpler by starting with Clebsch's theorem (Love 1927). For a sandwich plate in plane strain, the theorem takes the following form for any two distinct modes, here distinguished by the subscripts m and n:

$$\int_0^\ell \int_{-h_1}^{h_1} \rho_1 (U_{1m} U_{1n} + W_{1m} W_{1n}) \; dx \; dz$$

$$+ \int_0^\ell \int_{-h}^{-h_1} \rho_2 (U_{2m} U_{2n} + W_{2m} W_{2n}) \; dx \; dz$$

$$+ \int_0^\ell \int_{h_1}^{h} \rho_2 (U_{3m} U_{3n} + W_{3m} W_{3n}) \; dx \; dz = 0 \qquad (\omega_m \neq \omega_n). \quad (6.3.5)$$

The mode shapes may be expressed in terms of either the displacements or the plate displacements. According to the assumed form of the displacements in the general theory, the two sets of mode shapes are related to each other by the expressions

$$U_{1n} = z\Psi_{1n}$$
$$U_{2n} = -h_1(\Psi_{1n} - \Psi_{2n}) + z\Psi_{2n}$$
$$U_{3n} = h_1(\Psi_{1n} - \Psi_{2n}) + z\Psi_{2n}$$
$$W_{1n} = W_{2n} = W_{3n} = W_n$$

and by similar expressions with n replaced by m. When these are substituted in Eq. (6.3.5) and integration is carried out with respect to z, we fnd

$$\int_0^\ell \left[\left(\frac{4\rho_1 h_1}{3} + \rho_2 h_2 \right) h_1^2 \Psi_{1m} \Psi_{1n} + \right.$$

$$+ 3\rho_2 h_2 \left(h_1 \Psi_{1m} + \frac{2h_2 \Psi_{2m}}{3} \right) \left(h_1 \Psi_{1n} + \frac{2h_2 \Psi_{2n}}{3} \right)$$

$$\left. + 4(\rho_1 h_1 + \rho_2 h_2) W_m W_n \right] dx = 0 \qquad (\text{for } \omega_m \neq \omega_n), \quad (6.3.6)$$

which is the desired orthogonality conditon.

6.3.3   Transformation of Equations

In the manner of Mindline and Goodman (1950), we adopt the following transformation of the dependent variables:

$$\psi_1(x, t) = \phi_1(x, t) + \sum_{i=1}^{6} g_{1i}(x) \; f_i(t)$$

$$\psi_2(x, t) = \phi_2(x, t) + \sum_{i=1}^{6} g_{2i}(x) \; f_i(t) \qquad (6.3.7)$$

$$w(x, t) = \phi_3(x, t) + \sum_{i=1}^{6} g_{3i}(x) \; f_i(t)$$

In terms of the new variables ϕ_1, ϕ_2, and ϕ_3, the equations of motion (6.3.1) become

$$(2D_1 + E_2 h_1^2)\phi_1'' - 2G_1(\phi_1 + \phi_3') + 6G_2\frac{h_1}{h_2}(\phi_2 + \phi_3')$$

$$+ \sum_{i=1}^{6}\left[(2D_1 + 2E_2 h_1^2)g_{1i}'' - 2G_1(g_{1i} + g_{3i}') + 6G_2\frac{h_1}{h_2}(g_{2i} + g_{3i}')\right] f_i$$

$$-h_1(p_x^+ - p_x^-) = \left(\frac{4\rho_1 h_1^3}{3} + \rho_2 h_1^2 h_2\right)\ddot{\phi}_1 + \sum_{i=1}^{6}\left(\frac{4\rho_1 h_1^3}{3} + \rho_2 h_1^2 h_2\right) g_{1i}\ddot{f}_i$$

$$E_2 h_1 h_2 \phi_1'' + E_2\frac{2h_2^2}{3}\phi_2'' - 2G_2(\phi_2 + \phi_3') \tag{6.3.8}$$

$$+ \sum_{i=1}^{6}\left[E_2 h_1 h_2 g_{1i}'' + E_2\frac{2h_2^2}{3}g_{2i}'' - 2G_2(g_{2i} + g_{3i}')\right] f_i$$

$$+h_2(p_x^+ - p_x^-) = \rho_2 h_2^2\left(h_1\ddot{\phi}_1 + \frac{2h_2\ddot{\phi}_2}{3}\right) + \sum_{i=1}^{6}\rho_2 h_2^2\left(h_1 g_{1i} + \frac{2h_2 g_{2i}}{3}\right)\ddot{f}_i$$

$$G_1(\phi_1' + \phi_3'') + 2G_2(\phi_2' + \phi_3'') + \sum_{i=1}^{6}[G_1(g_{1i}' + g_{3i}'') + 2G_2(g_{2i}' + g_{3i}'')]f_i$$

$$+p_z^+ + p_z^- = 2(\rho_1 h_1 + \rho_2 h_2)\ddot{\phi}_3 + \sum_{i=1}^{6} 2(\rho_1 h_1 + \rho_2 h_2)g_{3i}\ddot{f}_i.$$

Similarly, the time-dependent boundary conditions in Eqs. (6.3.2) become

$$B_i(\phi_1, \phi_2, \phi_3) + \sum_{i=1}^{6} B_i(g_{1j}, g_{2j}, g_{3j})f_j(t) = f_i(t) \qquad (i = 1, 2, \ldots, 6).$$

$$(6.3.9)$$

The purpose of the transformation is to render the boundary conditions on ϕ_1, ϕ_2, and ϕ_3 homogeneous and independent of time, that is, to make

$$B_i(\phi_1, \phi_2, \phi_3) = 0 \qquad (i = 1, 2, \ldots, 6). \tag{6.3.10}$$

Thus, according to Eqs. (6.3.9), we must have

$$\sum_{i=1}^{6} B_i(g_{1j}, g_{2j}, g_{3j})f_j(t) = f_i(t) \qquad (i = 1, 2, \ldots, 6).$$

These may be satisfied simply by requiring

$$\begin{aligned} B_i(g_{1j}, g_{2j}, g_{3j}) &= 0 \quad \text{if } i \neq j \quad (i, j = 1, 2, \ldots, 6) \\ &= 1 \quad \text{if } i = j \quad (i, j = 1, 2, \ldots, 6), \end{aligned} \tag{6.3.11}$$

from which the functions g_{1j}, g_{2j}, and g_{3j} then may be determined by Mindlin and Goodman's procedure.

Finally, the initial conditions for ϕ_1, ϕ_2, and ϕ_3 require that, on account of Eqs. (6.3.3),

$$\phi_1(x, 0) = \psi_{10}(x) - \sum_{i=1}^{6} g_{1i}(x) f_i(0), \quad \phi_2(x, 0) = \ldots, \quad \phi_3(x, 0) = \ldots$$

$$(6.3.12)$$

$$\dot{\phi}_1(x, 0) = \dot{\psi}_{10}(x) - \sum_{i=1}^{6} g_{1i}(x) \dot{f}_i(0), \quad \dot{\phi}_2(x, 0) = \ldots, \quad \dot{\phi}_3(x, 0) = \ldots$$

6.3.4 Solution of Equations

Since the boundary conditions expressed in the new dependent variables are no longer time-dependent, the classical method of separation of variables may now be applied by taking the variables in the series form

$$[\phi_1(x, t), \phi_2(x, t), \phi_3(x, t)] = \sum_{n=1}^{\infty} [\Psi_{1n}(x), \Psi_{2n}(x), W_n(x)] T_n(t). \quad (6.3.13)$$

Again, as was mentioned in Section 6.1 in the simply supported case, it is understood that the principal modes $(\Psi_{1n}, \Psi_{2n}, W_n)$, assumed to be known, may actually be triply infinite in number, even though only a single infinite series is written out in Eqs. (6.3.13). Hereafter, the same interpretation is implied whenever such infinite series expansions are used. The major task in the classical method is to determine the time function $T_n(t)$ in the above series.

The procedure for determining $T_n(t)$ is standard. Details were given in the original paper (Yu 1960), and only an outline is described here. Equations (6.3.13) first are substituted into the governing Eqs. (6.3.8), which are subsequently modified through the use of Eqs. (6.3.4). Those terms in the results that are not yet in terms of the principal modes $(\Psi_{1n}, \Psi_{2n}, W_n)$ are next expanded in the form of infinite series by means of the orthogonality condition in Eqs. (6.3.6). By equating the nth terms in the results, we find the governing equation

$$\ddot{T}_n(t) + \omega_n^2 T_n(t) = Q_n(t) - \sum_{i=1}^{6} G_{in} \ddot{f}_i(t) + \sum_{i=1}^{6} G_{in}^* f_i(t), \quad (6.3.14)$$

where

$$Q_n(t) = \frac{2}{I_{nn}} \int_0^\ell (p_x h_1 \Psi_{1n} + p_x h_2 \Psi_{2n} + p_z W_n) \, dx,$$

$$G_{in} = \frac{2}{I_{nn}} \int_0^\ell \left\{ \left[\frac{2\rho_1 h_1^3 g_{1i}}{3} + \rho_2 h_1 h_2 (2h_1 g_{1i} + h_2 g_{2i}) \right] \Psi_{1n} \right.$$

$$\left. + \rho_2 h_2^2 \left(h_1 g_{1i} + \tfrac{2}{3} h_2 g_{2i} \right) \Psi_{2n} + 2(\rho_1 h_1 + \rho_2 h_2) g_{3i} W_n \right\} \, dx,$$

$$G_{in}^* = \frac{2}{I_{nn}} \int_0^\ell \left\{ \left[(D_1 + 2E_2 h_1^2) g_{1i}'' + E_2 h_1 h_2 g_{2i}'' - G_1(g_{1i} + g_{3i}') \right] \Psi_{1n} \right.$$

$$+ \left[E_2 h_1 h_2 g_{1i}'' + \tfrac{2}{3} E_2 h_2^2 g_{2i}'' - 2G_2(g_{2i} + g_{3i}') \right] \Psi_{2n}$$

$$\left. + \left[G_1(g_{1i}' + g_{3i}'') + 2G_2(g_{2i}' + g_{3i}'') \right] W_n \right\} dx, \tag{6.3.15}$$

and

$$I_{nn} = \int_0^\ell \left[\left(\frac{4\rho_1 h_1^3}{3} + \rho_2 h_1^2 h_2 \right) \Psi_{1n}^2 + 3\rho_2 h_2 \left(h_1 \Psi_{1n} + \frac{2h_2 \Psi_{2n}}{3} \right)^2 \right.$$

$$\left. + 4(\rho_1 h_1 + \rho_2 h_2) W_n^2 \right] dx.$$

It is well known that the solution of Eq. (6.3.14) has the form

$$T_n(t) = A_n \cos \omega_n t + B_n \sin \omega_n t$$

$$+ \frac{1}{\omega_n} \int_0^t [Q_n(\tau) - \sum_{i=1}^6 G_{in} \ddot{f}_i(\tau) + \sum_{i=1}^6 G_{in}^* f_i(\tau)] \sin \omega_n(t - \tau) \, d\tau.$$

The last step involves the determination of A_n and B_n from the initial displacements and velocities given in Eqs. (6.3.3) by expanding the latter quantities in infinite series form in terms of the principal modes in the same manner as the quantities $Q_n(t)$, G_{in}, and G_{in}^* were treated, through the use of the condition of orthogonality. In terms of the initial displacements and velocities, the solution becomes finally

$$T_n(t) = C_n \cos \omega_n t + \frac{1}{\omega_n} D_n \sin \omega_n t - \sum_{i=1}^6 G_{in} f_i(t)$$

$$+ \frac{1}{\omega_n} \int_0^t \left[Q_n(\tau) + \sum_{i=1}^6 (\omega_n^2 G_{in} + G_{in}^*) f_i(\tau) \right] \sin \omega_n(t - \tau) \, d\tau, \tag{6.3.16}$$

where we now have, instead of A_n and B_n,

$$C_n = \frac{2}{I_{nn}} \int_0^\ell \left\{ \left[\left(\frac{2\rho_1 h_1}{3} + 2\rho_2 h_2 \right) h_1^2 \psi_{10} + \rho_2 h_1 h_2^2 \psi_{20} \right] \Psi_{1n} \right.$$

$$\left. + \rho_2 h_2^2 \left(h_1 \psi_{10} + \tfrac{2}{3} h_2 \psi_{20} \right) \Psi_{2n} + 2(\rho_1 h_1 + \rho_2 h_2) w_0 W_n \right\} dx \tag{6.3.17}$$

$$D_n = \frac{2}{I_{nn}} \int_0^\ell \left\{ \left[\left(\frac{2\rho_1 h_1}{3} + 2\rho_2 h_2 \right) h_1^2 \dot{\psi}_{10} + \rho_2 h_1 h_2^2 \dot{\psi}_{20} \right] \Psi_{1n} \right.$$

$$\left. + \rho_2 h_2^2 \left(h_1 \dot{\psi}_{10} + \tfrac{2}{3} h_2 \dot{\psi}_{20} \right) \Psi_{2n} + 2(\rho_1 h_1 + \rho_2 h_2) \dot{w}_0 W_n \right\} dx.$$

6.3.5 An Example

Let us consider a sandwich plate in plane strain, simply supported at the edges $x = 0$ and $x = \ell$. A transverse deflection varying with time is prescribed according to a given function $f_6(t)$ at $x = \ell$. The boundary conditions are thus

$$(D_1 + 2E_2h_1^2)\psi_1'(0, t) + E_2h_1h_2\psi_2'(0, t) = 0$$

$$(D_1 + 2E_2h_1^2)\psi_1'(\ell, t) + E_2h_1h_2\psi_2'(\ell, t) = 0$$

$$E_2h_1h_2\psi_1'(0, t) + \frac{2E_2h_2^2\psi_2'(0, t)}{3} = 0$$

$$E_2h_1h_2\psi_1'(\ell, t) + \frac{2E_2h_2^2\psi_2'(\ell, t)}{3} = 0 \qquad (6.3.18)$$

$$w(0, t) = 0$$

$$w(\ell, t) = f_6(t).$$

Since the functions $f_i(t) = 0$ with $i = 1, 2, 3, 4, 5$, we have

$$g_{1i} = g_{2i} = g_{3i} = 0 \qquad (i = 1, 2, 3, 4, 5). \qquad (6.3.19)$$

On the other hand, since the prescribed function $f_6(t)$ is not 0, the corresponding functions g_{16}, g_{26}, and g_{36} are determined by Eqs. (6.3.11), which in this example take the form, according to Eqs. (6.3.18),

$$(D_1 + 2E_2h)_1^2)g_{16}'(0) + E_2h_1h_2g_{26}'(0) = 0$$

$$(D_1 + 2E_2h_1^2)g_{16}'(\ell) + E_2h_1h_2g_{26}'(\ell) = 0$$

$$E_2h_1h_2g_{16}'(0) + \frac{2E_2h_2^2g_{26}'(0)}{3} = 0$$

$$E_2h_1h_2g_{16}'(\ell) + \frac{2E_2h_2^2g_{26}'(\ell)}{3} = 0$$

$$g_{36}(0) = 0$$

$$g_{36}(\ell) = 1$$

These conditions are satisfied by taking

$$g_{16}(x) = g_{26}(x) = 0 \qquad g_{36}(x) = \frac{x}{\ell}. \qquad (6.3.20)$$

In a simply supported sandwich plate in plane strain, there is a triply infinite set of principal modes with the following form (Yu 1960):

$$\Psi_{1n} = a_n \cos \frac{n\pi x}{\ell}$$

$$\Psi_{2n} = b_n \cos \frac{n\pi x}{\ell} \qquad (n = 1, 2, \ldots) \qquad (6.3.21)$$

$$W_n = c_n \sin \frac{n\pi x}{\ell}.$$

For each value of n, there are three sets of natural frequencies and values of (a_n, b_n, c_n) defining the mode shape.

We are now in a position to evaluate the time function $T_n(t)$ in Eq. (6.3.16). For simplicity, let us take all initial displacements and velocities equal to zero, which yield $C_n = 0$ and $D_n = 0$ according to Eqs. (6.3.17). In the absence of the surface tractions p_x and p_z, we have $Q_n = 0$ according to the first of Eqs. (6.3.15). Finally, according to the second and third of Eqs. (6.3.15), and by substituting g_{1i}, g_{2i}, g_{3i}, with $i = 1, 2, \ldots, 6$ from Eqs. (6.3.19) and (6.3.20) and Ψ_{1n}, ψ_{2n}, W_n from Eqs. (6.3.21), we find

$$G_{in}^* = 0 \qquad (i = 1, 2, \ldots, 6)$$
$$G_{in} = 0 \qquad (i = 1, 2, \ldots, 5)$$

$$G_{6n} = 8(-1)^{n+1}(\rho_1 h_1 + \rho_2 h_2)\frac{c_n}{n\pi}\left[\left(\frac{4\rho_1 h_1^3}{3} + \rho_2 h_1^2 h_2\right) a_n^2\right.$$
$$\left. + 3\rho_2 h_2 \left(h_1 a_n + \frac{2h_2 b_n}{3}\right)^2 + 4(\rho_1 h_1 + \rho_2 h_2)c_n^2\right].$$

The time function is therefore

$$T_n(t) = G_{6n}\omega_n \int_0^t f_6(\tau) \sin \omega_n(t - \tau) \, d\tau - G_{6n} f_6(t). \qquad (6.3.22)$$

As a special case of this example, we take the simple form

$$f_6(t) = K \sin \omega t.$$

Equation (6.3.22) then yields

$$T_n(t) = \frac{KG_{6n}\omega^2}{\omega^2 - \omega_n^2}\left(\frac{\omega_n}{\omega} \sin \omega_n t - \sin \omega t\right).$$

The displacement w is, for instance, according to Eqs. (6.3.7) and (6.3.13),

$$w(x, t) = \sum_{n=1}^{\infty} c_n \sin \frac{n\pi x}{\ell} \frac{KG_{6n}\omega^2}{\omega^2 - \omega_n^2}\left(\frac{\omega_n}{\omega} \sin \omega_n t - \sin \omega t\right)$$
$$+ \frac{x}{\ell} K \sin \omega t,$$

from which

$$w(0, t) = 0$$
$$w(\ell, t) = K \sin \omega t = f_6(t)$$
$$w(x, 0) = 0$$
$$\dot{w}(x, 0) = K\omega \left[-\sum_{n=1}^{\infty} G_{6n} c_n \sin \frac{n\pi x}{\ell} + \frac{x}{\ell}\right] = 0.$$

The bracket in the last equation is zero because the summation is exactly the series expansion of the function x/ℓ. All boundary and initial conditions thus are satisfied.

References

Love, A.E.H. (1927) *A Treatise on the Mathematical Theory of Elasticity*, 4th Ed., p. 180. Cambridge University Press, Cambridge, England.

Mindlin, R.D. and L.E. Goodman. (1950) Beam Vibrations with Time-Dependent Boundary Conditions. *Journal of Applied Mechanics*, Vol. 17, pp. 377–380.

Ren, N. and Y.Y. Yu. (1967) Flexural and Extensional Vibrations of Two-Layered Plates. *AIAA Journal*, Vol. 5, pp. 797–799.

Timoshenko, S. (1955) *Vibration Problems in Engineering*, 3rd Ed. Van Nostrand, New York.

Yu, Y.Y. (1960) Forced Flexural Vibrations of Sandwich Plates in Plane Strain. *Journal of Applied Mechanics*, Vol. 27, pp. 535–540.

Yu. Y.Y. (1962) Damping of Flexural Vibrations of Sandwich Plates. *Journal of Aerospace Sciences*, Vol. 29, pp. 790–803.

Yu, Y.Y. and J. Lai. (1967) Application of Galerkin's Method to the Dynamic Analysis of Structures. *AIAA Journal*, Vol. 5, pp 792–795.

Yu, Y.Y. and N. Ren. (1965) Damping Parameters of Layered Plates and Shells. *Acoustical Fatigue in Aerospace Structures*, pp. 555–584. Syracuse University Press, Syracuse, New York.

7

Nonlinear Modeling for Large Deflections of Beams, Plates, and Shallow Shells

In the remaining chapters in this book, we shall extend our treatment to nonlinear vibrations of thin structures that involve large deflections. Nonlinear dynamical modeling for such structures is introduced in the present chapter. This is followed in Chapter 8 with discussions of nonlinear vibrations of layered beams and plates, again including both sandwiches and laminated composites. Chaotic vibrations of elastic beams are then explored in Chapter 9. Finally, in Chapter 10, nonlinear dynamical modeling for large deflections of piezoelectric plates is treated.

In developing a strategy for nonlinear dynamical modeling, we have restudied the works of previous authors, including particularly the equations of some of the most distinguished masters in the field. Since these equations are well accepted, they have provided not only renewed insight, but also reassurance for new extensions. In this chapter, we report on our experience in paying such a revisit to the famous equations of Timoshenko (1921), von Kármán (1910), and Marguerre (1938). The original Timoshenko beam equations were linear. Von Kármán's equations for large deflections of plates have been a standard topic covered in many books, such as those by Timoshenko (1940), Novozhilov (1948), Langhaar (1962), Fung (1965), Washizu (1968), Dym and Shames (1973), and Chia (1980). Attempts to incorporate the transverse shear effect into the von Kármán equations were made first by Eringen (1955) and then by Medwadowski (1958) and Reddy (1984). Marguerre's equations for large deflections of shallow shells may be found in the books by Washizu (1968) and Chia (1980).

Thus, in the first section in this chapter, we derive the equations for large deflections of a buckled Timoshenko beam (Yu 1992a). In the next two sections, we rederive von Kármán's equations for large deflections of a flat plate and Marguerre's equations for a shallow shell, which we shall further extend by incorporating the transverse shear effect (Yu 1991). All derivations are based on the use of a pseudo-variational equation of motion in nonlinear elasticity that was described in Chapter 1. In the last section in this chapter, some general remarks are offered on the use of the variational equations of motion based on our early and recent experiences (Yu 1992b, 1995a,b).

7.1   Equations for Large Deflections of a Buckled Timoshenko Beam

As was pointed out earlier, the transverse shear effect was first incorporated into the beam equations by Timoshenko (1921). In this section, we further consider the effects of large deflection and buckling in a Timoshenko beam (Yu 1992a).

Consider a rectangular beam with an axis in the x-direction and with the ends at $x = 0, \ell$. For convenience, the width of the beam in the y-direction is taken as unity, so that the treatment also may be applicable to a plate in plane strain. The middle plane of the beam thus is taken to be the xy-plane. In the z-direction, the beam extends from $z = -h$ to $z = h$.

7.1.1   Stress Equations of Motion

We take the three-dimensional displacements in the form

$$u_x = u + z\psi, \quad u_y = 0, \quad u_z = w \tag{7.1.1}$$

and the rotations in the form

$$\omega_{xy} = 0, \quad \omega_{yz} = 0, \quad \omega_{zx} = -\frac{\partial w}{\partial x}, \tag{7.1.2}$$

in which u, w, and ψ are the beam displacements as functions of x and t. Ignoring body forces, we substitute Eqs. (7.1.1) and (7.1.2) into the pseudo-variational equation (1.4.8) and carry out integration with respect to z over the thickness of the beam. There results

$$\int_{t_0}^{t_1} \int_0^\ell \left\{ \left[\frac{\partial N}{\partial x} - 2\rho h \ddot{u} \right] \delta u + \left[\frac{\partial}{\partial x} \left(Q + N \frac{\partial w}{\partial x} \right) + p_z - 2\rho h \ddot{w} \right] \delta w \right.$$

$$\left. + \left[\frac{\partial M}{\partial x} - Q - \frac{2\rho h^3}{3} \ddot{\psi} \right] \delta \psi \right\} \, dx \, dt$$

$$- \int_{t_0}^{t_1} \left\{ \left[N - p_x^{(0)} \right] \delta u + \left[Q + N \frac{\partial w}{\partial x} - p_z^{(0)} \right] \delta w \right.$$

$$\left. + \left[M - p_x^{(1)} \right] \delta \psi \right\}_{x=0}^{x=\ell} dt = 0, \tag{7.1.3}$$

where the lateral load $p_z = p_z^+ + p_z^-$ includes tractions at both the top and bottom of the beam or plate in the z-direction. In addition, the beam stresses and integrated tractions at the ends of the beam are defined by, respectively,

$$(N, Q, M) = \int_{-h}^{h} (\sigma_{xx}, \sigma_{zx}, \sigma_{xx}z)\ dz$$

$$\left(p_x^{(0)}, p_z^{(0)}, p_x^{(1)}\right) = \int_{-h}^{h} (p_x, p_z, p_x z)\ dx,$$

(7.1.4)

where N, Q, and M are the usual axial membrane force, shearing force, and bending moment, respectively, and the superscripts (0) and (1) denote edge tractions of zeroth and first orders, respectively.

Equation (7.1.3) is the pseudo-variational equation of motion of the beam, from which we obtain the stress equations of motion

$$N' - 2\rho h \ddot{u} = 0$$
$$(Q + Nw')' + p_z - 2\rho h \ddot{w} = 0 \qquad (0 \le x \le \ell)$$
$$M' - Q - \frac{2\rho h^3}{3} \ddot{\psi} = 0$$

(7.1.5)

and the boundary conditions

$$N = p_x^{(0)} \quad \text{or} \quad u = \text{prescribed}$$
$$Q + Nw' = p_z^{(0)} \quad \text{or} \quad w = \text{prescribed} \qquad (x = 0, \ell)$$
$$M = p_x^{(1)} \quad \text{or} \quad \psi = \text{prescribed}.$$

(7.1.6)

A prime now denotes differentiation with respect to x.

7.1.2   Displacement Equations of Motion

To develop relations between the beam stresses and beam displacements, we note again that, in nonlinear elasticity, the three-dimensional stresses are components of Kirchhoff's stress tensor. These are associated with components of Green's nonlinear strain tensor through the strain energy function in a similar manner as linear stresses and strains are related to each other in linear elasticity. Thus, we can write the nonlinear stress–strain relations

$$\sigma_{xx} = E\,\epsilon_{xx}, \qquad \sigma_{xx} = \frac{E}{1 - \nu^2}\epsilon_{xx}$$

(7.1.7)

for a beam and for a plate in plane strain, respectively. In the sequel, only the latter will be written out, and the former can be deduced simply by replacing the parenthesis $(1 - \nu^2)$ by unity. In the simplified nonlinear cases, the only nonlinear normal strain that appears is, by virtue of Eqs. (7.1.1) and (7.1.2),

$$\epsilon_{xx} = u' + \tfrac{1}{2}(w')^2.$$

Substitution of ϵ_{xx} in the second of Eqs. (7.1.7) and the result in Eqs. (7.1.4) yields

$$N = 2K\left[u' + \tfrac{1}{2}(w')^2\right] \qquad \text{with} \quad K = \frac{Eh}{1 - \nu^2}. \qquad (7.1.8)$$

Relations for M and Q have the same form as in the linear case:

$$M = \frac{E}{1 - \nu^2}\,\frac{2h^3}{3}\,\psi', \qquad Q = 2\kappa Gh(\psi + w'), \qquad (7.1.9)$$

where κ is the shear correction factor introduced in the manner of Mindlin (1951) and has the value $\pi^2/12$.

To concentrate on the lateral motion of a beam, we keep only the inertia term in the second of Eqs. (7.1.5) and ignore the other inertia terms in these equations. According to the simplified form of the first of Eqs. (7.1.5), the axial force N is then independent of x and uniform throughout the entire length of the beam. Simple integration of N over the length of the beam further yields

$$N = K\,\frac{2}{\ell}\left[u_0 + \frac{1}{2}\int_0^\ell (w')^2\,dx\right], \qquad (7.1.10)$$

where u_0 is the relative axial displacement between the two ends of the beam. By virtue of Eqs. (7.1.9) and (7.1.10), the second and the simplified form of the third of Eqs. (7.1.5) become

$$\frac{2Kh^2}{3}\,\psi''' + K\,\frac{2}{\ell}\left[u_0 + \frac{1}{2}\int_0^\ell (w')^2\,dx\right]w'' + p_z = 2\rho h\,\ddot{w} \qquad (7.1.11)$$

$$2\kappa Gh(\psi + w') = \frac{2Kh^2}{3}\,\psi'', \qquad (7.1.12)$$

which are the displacement equations of motion.

7.1.3 Dynamical Model

Consider a simply supported buckled beam under a lateral load

$$p_z = P_z \sin\frac{\pi x}{\ell}\,\cos\omega t,$$

where P_z is the amplitude of p_z and ω is the forcing frequency. A simple dynamical model may be constructed by adopting a single-mode approximation represented by

$$w(x,t) = W\,\tau(t)\,\sin\frac{\pi x}{\ell}, \qquad \psi(x,t) = \Psi\,\tau(t)\,\cos\frac{\pi x}{\ell}. \qquad (7.1.13)$$

Equation (7.1.12) then gives

$$\Psi = -\frac{1}{1 + \dfrac{K}{3\kappa Gh}\dfrac{\pi^2 h^2}{\ell^2}}\,\frac{\pi W}{\ell} = -\beta\,\frac{\pi W}{\ell},$$

and Eq. (7.1.11) becomes

$$2\rho h\, W\ddot{\tau} + 2K \left(\beta \frac{\pi^4 h^2}{3\ell^4} + \frac{\pi^2 u_0}{\ell^3} \right) W\tau + \frac{1}{2} K \frac{\pi^4}{\ell^4} W^3 \tau^3 = P_z \cos \omega t, \quad (7.1.14)$$

where a dimensionless parameter β has been introduced.

To incorporate buckling explicitly into the formulation, we solve Eq. (7.1.14) for the linear natural frequency by ignoring the nonlinear term:

$$\omega_1^2 = \frac{K}{\rho h} \left(\beta \frac{\pi^4 h^2}{3\ell^4} + \frac{\pi^2 u_0}{\ell^3} \right).$$

At inception of buckling, ω_1 becomes 0. The value of u_0 derived for this situation will be denoted by u_{cr}, given by

$$\frac{u_{cr}}{\ell} = -\beta \frac{\pi^2 h^2}{3\ell^2}.$$

We next introduce the ratio $r = u_0/u_{cr}$, which also may be interpreted as the ratio of the applied axial load to the critical buckling load. In terms of r, Eq. (7.1.14) takes the final form

$$2\rho h\, W\ddot{\tau} + 2K h^2 \frac{\pi^4}{\ell^4} \frac{\beta}{3} (1-r) W\tau + \frac{\pi^4}{2\ell^4} K W^3 \tau^3 = P_z \cos \omega t. \quad (7.1.15)$$

It is interesting to note that, for incipient buckling for which $r = 1$, the second term drops out and the parameter β also disappears from the equation. This particular situation then is no longer dependent upon the transverse shear effect, since this effect is imbedded in the parameter β only.

The dynamical model presented in Eq. (7.1.15) is based on the Timoshenko beam theory. The transverse shear effect may be easily suppressed by letting the shear correction factor κ be equal to infinity. This renders $\beta = 1$, which is the value to be used if the classical beam theory is adopted.

Equation (7.1.15) has the form

$$\ddot{x} + ax + x^3 = B \cos t, \quad (7.1.16)$$

which is a Duffing's equation without damping. Equations of similar form also have been adopted in nonlinear vibration analysis of elastic plates and shells of both homogeneous and layered types.

7.2   Von Kármán Equations for Large Deflections of a Plate: Incorporation of Transverse Shear Effect

The elastic isotropic plate in Figure 3.1.1 is again considered here. The classical linear equations of an isotropic plate for flexure and extension as discussed in

Sections 3.1 and 3.3 are uncoupled from each other. For large deflections, the plate equations are nonlinear, and flexure and extension become coupled. The results are the famous von Kármán Equations (1910). Our derivation (Yu 1991) makes use of one of the two pseudo-variational equations of motion discussed earlier. Nonlinearity appears not only in the stress equations of motion and traction boundary conditions, but also in the stress–strain–displacement relations.

7.2.1 *Variational Equation of Motion*

To derive the stress equations of motion and traction boundary conditions, we make use of the pseudo-variational equation (1.4.8) for the new simplified nonlinear case of large deformations. Anticipating coupling between flexure and extension, we assume the displacements in the form

$$u_x(x, y, z, t) = u - z\,\frac{\partial w}{\partial x}, \qquad u_y(x, y, z, t) = v - z\,\frac{\partial w}{\partial y} \tag{7.2.1}$$

$$u_z(x, y, z, t) = w.$$

Of the three rotations, ω_{xy} is about the z axis that is normal to the plane of the plate, and ω_{yz} and ω_{zx} are about the x- and y-axes that are in the plane of the plate. The former is assumed to be much smaller than the latter and will be neglected. Upon substitutions of Eqs. (7.2.1) into (1.1.7), the latter two rotations become

$$\omega_{zx} = -\frac{\partial w}{\partial x}, \qquad \omega_{yz} = \frac{\partial w}{\partial y}. \tag{7.2.2}$$

By substituting Eqs. (7.2.1) and (7.2.2) into the pseudo-variational equation (1.4.8) and carrying out integration with respect to z over the thickness of the plate, the final result has the form

$$
\int_{t_0}^{t_1} dt \int\!\int_A \left\{ \left[\frac{\partial N_x}{\partial x} + \frac{\partial N_{xy}}{\partial y} + P_x^{(0)} + f_x^{(0)} - 2\rho h \ddot{u} \right] \delta u \right.
$$

$$
+ \left[\frac{\partial N_{xy}}{\partial x} + \frac{\partial N_y}{\partial y} + P_y^{(0)} + f_y^{(0)} - 2\rho h \ddot{v} \right] \delta v
$$

$$
+ \left[\frac{\partial^2 M_x}{\partial x^2} + 2\frac{\partial^2 M_{xy}}{\partial x \partial y} + \frac{\partial^2 M_y}{\partial y^2} \right.
$$

$$
+ \frac{\partial}{\partial x}\left(N_x \frac{\partial w}{\partial x} + N_{xy}\frac{\partial w}{\partial y} \right) + \frac{\partial}{\partial y}\left(N_{xy}\frac{\partial w}{\partial x} + N_y \frac{\partial w}{\partial y} \right)
$$

$$
+ \frac{\partial P_x^{(1)}}{\partial x} + \frac{\partial P_y^{(1)}}{\partial y} + P_z^{(0)}
$$

$$
+ \frac{\partial f_x^{(1)}}{\partial x} + \frac{\partial f_y^{(1)}}{\partial y} + f_z^{(0)}
$$

$$
\left. \left. - 2\rho h \ddot{w} + \tfrac{2}{3}\rho h^3 \nabla^2 \ddot{w} \right] \delta w \right\} dx\, dy
$$

$$
-\int_{t_0}^{t_1} dt \oint_{C_p} \left\{ \left[N_n - p_n^{(0)} \right] \delta u_n + \left[N_{ns} - p_s^{(0)} \right] \delta u_s \right.
$$

$$
+ \left[M_n - p_n^{(1)} \right] \delta \left(\frac{\partial w}{\partial n} \right) + \left[\frac{\partial M_n}{\partial n} + 2 \frac{\partial M_{ns}}{\partial s} \right.
$$

$$
+ N_n \frac{\partial w}{\partial n} + N_{ns} \frac{\partial w}{\partial s} + P_n^{(1)} + f_n^{(1)} - p_z^{(0)}
$$

$$
\left. \left. - \frac{\partial p_s^{(1)}}{\partial s} + \frac{2}{3} \rho h^3 \frac{\partial \ddot{w}}{\partial n} \right] \delta w \right\} \ ds = 0. \tag{7.2.3}
$$

Equation (7.2.3) is the two-dimensional pseudo-variational equation of motion for a plate with large deflections, from which the stress equations of motion and traction boundary conditions may be written immediately. The stress equations of motion include many more terms than the original von Kármán equations. These are terms involving inertia forces, body forces, and surface tractions in all three directions. To be able to accommodate applied forces and tractions in all directions can be significant from an engineering and physical standpoint. For instance, due to coupling between flexure and extension, tractions in the tangential directions can excite a motion in the transverse direction even in an isotropic plate.

The usual form of the von Kármán equations is for the static case without body forces and tractions in the x- and y-directions and with only lateral load in the z-direction. In this case, we obtain from the first two lines of Eq. (7.2.3)

$$
\frac{\partial N_x}{\partial x} + \frac{\partial N_{xy}}{\partial y} = 0, \qquad \frac{\partial N_{xy}}{\partial x} + \frac{\partial N_y}{\partial y} = 0 \qquad \text{(in } A\text{)},
$$

and from the remaining lines in the double integral

$$
\frac{\partial^2 M_x}{\partial x^2} + 2 \frac{\partial^2 M_{xy}}{\partial x \partial y} + \frac{\partial^2 M_y}{\partial y^2} + N_x \frac{\partial^2 w}{\partial x^2}
$$

$$
+ 2 N_{xy} \frac{\partial^2 w}{\partial x \partial y} + N_y \frac{\partial^2 w}{\partial y^2} + P_z^{(0)} = 0 \qquad \text{(in } A\text{)}. \tag{7.2.4}
$$

The boundary conditions derived from the contour integral are

$$
N_n - p_n^{(0)} = 0 \quad \text{or} \quad u_n = \text{prescribed}
$$

$$
N_{ns} - p_s^{(0)} = 0 \quad \text{or} \quad u_s = \text{prescribed}
$$

$$
M_n - p_n^{(1)} = 0 \quad \text{or} \quad \frac{\partial w}{\partial n} = \text{prescribed} \tag{7.2.5}
$$

$$
\frac{\partial M_n}{\partial n} + 2 \frac{\partial M_{ns}}{\partial s} + N_n \frac{\partial w}{\partial n} + N_{ns} \frac{\partial w}{\partial s} - p_z^{(0)} = 0 \quad \text{or} \quad w = \text{prescribed}.
$$

As in the classical linear flexural theory, the transverse shearing forces Q_x and Q_y are not directly involved in the plate stress–strain–displacement relations since the corresponding plate transverse shear strains are 0, but they still can be expressed in terms of the moments and other quantities as was discussed in Section 3.1. In terms of the transverse shearing force, the last traction boundary condition from Eqs. (7.2.3) also can be written in the form

$$Q_n + \frac{\partial M_{ns}}{\partial s} + N_n \frac{\partial w}{\partial n} + N_{ns} \frac{\partial w}{\partial s} - p_z^{(0)} = 0.$$

7.2.2 Displacements, Strains, and Stresses

The displacements in Eqs. (7.2.1) are clearly a combination of Eqs. (3.1.1) for flexure and Eqs. (3.3.1) for extension. The corresponding linear strains are

$$e_{xx} = \frac{\partial u}{\partial x} - z\frac{\partial^2 w}{\partial x^2}, \qquad e_{yy} = \frac{\partial v}{\partial y} - z\frac{\partial^2 w}{\partial y^2}$$

$$e_{xy} = \frac{1}{2}\left(\frac{\partial u}{\partial y} + \frac{\partial v}{\partial x}\right) - z\frac{\partial^2 w}{\partial x \partial y}. \tag{7.2.6}$$

The nonlinear strains for the simplified case of large deformations as given by Eqs. (1.1.11) thus take the form

$$\epsilon_{xx} = \frac{\partial u}{\partial x} - z\frac{\partial^2 w}{\partial x^2} + \frac{1}{2}\left(\frac{\partial w}{\partial x}\right)^2$$

$$\epsilon_{yy} = \frac{\partial v}{\partial y} - z\frac{\partial^2 w}{\partial y^2} + \frac{1}{2}\left(\frac{\partial w}{\partial y}\right)^2 \tag{7.2.7}$$

$$\epsilon_{xy} = \frac{1}{2}\left(\frac{\partial u}{\partial y} + \frac{\partial v}{\partial x}\right) - z\frac{\partial^2 w}{\partial x \partial y} + \frac{1}{2}\frac{\partial w}{\partial x}\frac{\partial w}{\partial y}.$$

We note again that, in nonlinear elasticity theory, the components of Kirchhoff's stress tensor are associated with those of Green's nonlinear strain tensor in a similar manner as linear stresses and strains are related to each other in linear elasticity. Since the stress–strain relations in both linear flexure and extension were derived from the three-dimensional relations in Eq. (3.1.11), these still can be used as the starting point for the nonlinear case by simply replacing the linear stresses and strains by the corresponding nonlinear ones; thus,

$$\sigma_{xx} = E\frac{\epsilon_{xx} + \nu\epsilon_{yy}}{1 - \nu^2} + \frac{\sigma_{zz}\nu}{1 - \nu}$$

$$\sigma_{yy} = E\frac{\epsilon_{yy} + \nu\epsilon_{xx}}{1 - \nu^2} + \frac{\sigma_{zz}\nu}{1 - \nu}$$

$$\sigma_{xy} = 2\mu\epsilon_{xy} = \frac{E\epsilon_{xy}}{1 + \nu}.$$

By substituting these stresses into Eqs. (3.1.4) and (3.3.4), carrying out the integrations, and ignoring those involving σ_{zz} as before, there result

$$N_x = C\left[\frac{\partial u}{\partial x} + v\,\frac{\partial v}{\partial y} + \frac{1}{2}\left(\frac{\partial w}{\partial x}\right)^2 + \frac{1}{2}v\left(\frac{\partial w}{\partial y}\right)^2\right]$$

$$N_y = C\left[\frac{\partial v}{\partial y} + v\,\frac{\partial u}{\partial x} + \frac{1}{2}\left(\frac{\partial w}{\partial y}\right)^2 + \frac{1}{2}v\left(\frac{\partial w}{\partial x}\right)^2\right]$$

$$N_{xy} = \frac{1}{2}C(1-v)\left[\frac{\partial u}{\partial y} + \frac{\partial v}{\partial x} + \frac{\partial w}{\partial x}\,\frac{\partial w}{\partial y}\right]$$

$$M_x = -D\left(\frac{\partial^2 w}{\partial x^2} + v\,\frac{\partial^2 w}{\partial y^2}\right) \qquad (7.2.8)$$

$$M_y = -D\left(\frac{\partial^2 w}{\partial y^2} + v\,\frac{\partial^2 w}{\partial x^2}\right)$$

$$M_{xy} = -D(1-v)\frac{\partial^2 w}{\partial x \partial y}.$$

The linear parts of the plate stress–displacement relations of course remain the same as those for the linear cases of flexure and extension.

Substitution of the plate stress–displacement relations into the stress equations of motion yields the displacement equations of motion, which will not be written out here.

7.2.3 Incorporation of Transverse Shear Effect

We shall next generalize the above derivation by including the effect of transverse shear (Yu 1991). This is essentially to have a merger between the von Kármán and Mindlin equations. The displacements are now assumed as follows:

$$u_x(x, y, z, t) = u + z\psi, \quad u_y(x, y, z, t) = v + z\phi, \quad u_z(x, y, z, t) = w. \quad (7.2.9)$$

The rotations are still taken in the same form as in Eqs. (7.2.2).

The variational equation of motion for the plate is derived similarly as before, from Eq. (1.4.8). The result is

$$\int_{t_0}^{t_1} dt \int\!\!\int_A \left\{ \left[\frac{\partial N_x}{\partial x} + \frac{\partial N_{xy}}{\partial y} + P_x^{(0)} + f_x^{(0)} - 2\rho h\ddot{u}\right]\delta u \right.$$

$$+ \left[\frac{\partial N_{xy}}{\partial x} + \frac{\partial N_y}{\partial y} + P_x^{(0)} + f_y^{(0)} - 2\rho h\ddot{v}\right]\delta v$$

$$+ \left[\frac{\partial M_x}{\partial x} + \frac{\partial M_{xy}}{\partial y} - Q_x + P_x^{(1)} + f_x^{(1)} - \frac{2}{3}\rho h^3\ddot{\psi}\right]\delta\psi$$

$$+ \left[\frac{\partial M_{xy}}{\partial x} + \frac{\partial M_y}{\partial y} - Q_y + P_y^{(1)} + f_y^{(1)} - \frac{2}{3}\rho h^3\ddot{\phi}\right]\delta\phi$$

$$+ \left[\frac{\partial}{\partial x}\left(Q_x + N_x\frac{\partial w}{\partial x} + N_{xy}\frac{\partial w}{\partial y}\right)\right.$$

$$+ \frac{\partial}{\partial y}\left(Q_y + N_{xy}\frac{\partial w}{\partial x} + N_y\frac{\partial w}{\partial y}\right)$$

$$+ \left. P_z^{(0)} + f_z^{(0)} - 2\rho h\ddot{w}\right]\delta w\Big\}\ dx\ dy$$

$$- \int_{t_0}^{t_1} dt \oint_{C_p} \left\{ \left[N_n - p_n^{(0)}\right]\delta u_n + \left[N_{ns} - p_s^{(0)}\right]\delta u_s \right.$$

$$+ \left[M_n - p_n^{(1)}\right]\delta\psi_n + \left[M_{ns} - p_s^{(1)}\right]\delta\psi_s$$

$$+ \left.\left[Q_n + N_n\frac{\partial w}{\partial n} + N_{ns}\frac{\partial w}{\partial s}\right]\delta w\right\}ds = 0,$$

from which stress equations of motion and traction boundary conditions now may
be written in the usual manner.

The plate stress–displacement relations are also derived similarly. The results
are

$$N_x = C\left[\frac{\partial u}{\partial x} + v\frac{\partial v}{\partial y} + \frac{1}{2}\left(\frac{\partial w}{\partial x}\right)^2 + \frac{1}{2}v\left(\frac{\partial w}{\partial y}\right)^2\right]$$

$$N_y = C\left[\frac{\partial v}{\partial y} + v\frac{\partial u}{\partial x} + \frac{1}{2}\left(\frac{\partial w}{\partial y}\right)^2 + \frac{1}{2}v\left(\frac{\partial w}{\partial x}\right)^2\right]$$

$$N_{xy} = C(1 - v)\left[\frac{\partial v}{\partial x} + \frac{\partial u}{\partial y} + \frac{\partial w}{\partial x}\frac{\partial w}{\partial y}\right]$$

$$M_x = D\left(\frac{\partial\psi}{\partial x} + v\frac{\partial\phi}{\partial y}\right)$$

$$M_y = D\left(\frac{\partial\phi}{\partial y} + v\frac{\partial\psi}{\partial x}\right)$$

$$M_{xy} = D(1 - v)\left(\frac{\partial\phi}{\partial x} + \frac{\partial\psi}{\partial y}\right)$$

$$Q_x = 2\kappa\mu h\left(\psi + \frac{\partial w}{\partial x}\right)$$

$$Q_y = 2\kappa\mu h\left(\phi + \frac{\partial w}{\partial y}\right),$$

where

$$C = \frac{2Eh}{1 - v^2}, \qquad D = \frac{2Eh^3}{3(1 - v^2)},$$

and other notations are the same as before.

7.3 Marguerre Equations for Large Deflections of a Shallow Shell: Incorporation of Transverse Shear Effect

In this section, the Marguerre (1938) equations for large deflections of a shallow shell are generalized first by including the transverse shear effect. The generalized equations then reduce to the original Marquerre equations when the transverse shear effect is suppressed. The equations of a shallow shell also reduce to those of a flat plate and a beam as special cases. Our derivation is still based on the use of a pseudo-variational equation of motion (Yu 1991).

As is shown in Figure 7.3.1 for the xz-plane, the middle surface of the shell before deformation is defined by $w_0 = w_0(x, y)$. When $w_0 = 0$, the shallow shell becomes a flat plate. The coordinate ζ is normal to and measured from the middle surface of the shell. The top and bottom boundary surfaces of the shell are thus at $\zeta = h$ and $\zeta = -h$, respectively, with a plan area A. The shell also has a right cylindrical boundary surface S intersecting the middle surface along a closed contour C. The normal and tangential directions at a point on C are denoted by n and s, respectively.

7.3.1 Variational Equation of Motion

The displacements in the shallow shell are taken in the form

$$u_x = u + \zeta \, \psi, \qquad u_y = v + \zeta \, \phi, \qquad u_z = w \tag{7.3.1}$$

and the rotations in the form

$$\omega_{zx} = -\frac{\partial w_0}{\partial x} - \frac{\partial w}{\partial x}, \qquad \omega_{yz} = \frac{\partial w_0}{\partial y} + \frac{\partial w}{\partial y}. \tag{7.3.2}$$

By substituting these into the pseudo-variational Eq. (1.4.8) and carrying out integration with respect to ζ over the thickness of the shallow shell, there results

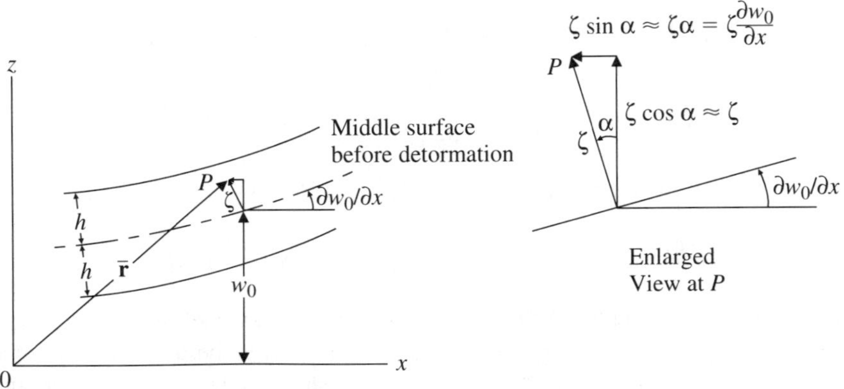

Fig. 7.3.1 Shallow shell before deformation.

$$\int_{t_0}^{t_1} dt \int_A \int \left(\left\{ \frac{\partial N_x}{\partial x} + \frac{\partial N_{xy}}{\partial y} + P_x^{(0)} + f_x^{(0)} - 2\rho h \ddot{u} \right\} \delta u \right.$$

$$+ \left\{ \frac{\partial N_{xy}}{\partial x} + \frac{\partial N_y}{\partial y} + P_x^{(0)} + f_y^{(0)} - 2\rho h \ddot{v} \right\} \delta v$$

$$+ \left\{ \frac{\partial M_x}{\partial x} + \frac{\partial M_{xy}}{\partial y} - Q_x + P_x^{(1)} + f_x^{(1)} - \frac{2}{3}\rho h^3 \ddot{\psi} \right\} \delta\psi$$

$$+ \left\{ \frac{\partial M_{xy}}{\partial x} + \frac{\partial M_y}{\partial y} - Q_y + P_y^{(1)} + f_y^{(1)} - \frac{2}{3}\rho h^3 \ddot{\phi} \right\} \delta\phi$$

$$+ \left\{ \frac{\partial}{\partial x}\left[Q_x + N_x \frac{\partial}{\partial x}(w_0 + w) + N_{xy}\frac{\partial}{\partial y}(w_0 + w) \right] \right.$$

$$+ \frac{\partial}{\partial y}\left[Q_y + N_{xy}\frac{\partial}{\partial x}(w_0 + w) + N_y\frac{\partial}{\partial y}(w_0 + w) \right]$$

$$\left. + P_z^{(0)} + f_z^{(0)} - 2\rho h \ddot{w} \right\} \delta w \Bigg) dx\, dy$$

$$- \int_{t_0}^{t_1} dt \oint_{C_p} \left\{ \left[N_n - p_n^{(0)} \right] \delta u_n + \left[N_{ns} - p_s^{(0)} \right] \delta u_s \right.$$

$$+ \left[M_n - p_n^{(1)} \right] \delta\psi_n + \left[M_{ns} - p_s^{(1)} \right] \delta\psi_s$$

$$+ \left[Q_n + N_n \frac{\partial(w_0 + w)}{\partial n} \right.$$

$$\left. + N_{ns}\frac{\partial(w_0 + w)}{\partial s} \right] \delta w \right\} ds = 0, \tag{7.3.3}$$

where the contour integral covers C_p, the part of the shell contour on which tractions are prescribed, and other notations are similar to those for a plate and are as follows:

$$(N_x, \ldots, Q_x, \ldots, M_x, \ldots) = \int_{-h}^{h} (\sigma_{xx}, \ldots, \sigma_{zx}, \ldots, \sigma_{xx}\zeta, \ldots)\, d\zeta$$

$$(f_x^{(0)}, \ldots, f_x^{(1)}, \ldots) = \int_{-h}^{h} (f_x, \ldots, f_x\zeta, \ldots)\, d\zeta$$

$$P_x^{(0)} = (p_x)_h + (p_x)_{-h}, \ldots, \qquad P_x^{(1)} = h[(p_x)_h - (p_x)_{-h}], \ldots$$

$$(p_n^{(0)}, \ldots, p_n^{(1)}, \ldots) = \int_{-h}^{h} (p_n, \ldots, p_n\zeta, \ldots)\, d\zeta.$$

These four lines define the shell stresses, shell body forces, shell surface tractions, and shell edge tractions, respectively. Equation (7.3.3) is the pseudo-variational equation for the shallow shell, from which the stress equations of motion and traction boundary conditions can now be written in the usual manner. The results are similar to those for a plate in that tractions applied to the top and bottom boundary surfaces in all directions have been accommodated.

7.3.2 Displacements, Strains, and Stresses

Consider a point P in the shallow shell in Figure 7.3.1 before deformation. As rotations are assumed to be much smaller than unity, the radius vector of P can be written as, according to the enlarged view at P in the figure,

$$\mathbf{r} = \mathbf{i}_x \left(x - \zeta \frac{\partial w_0}{\partial x} \right) + \mathbf{i}_y \left(y - \zeta \frac{\partial w_0}{\partial y} \right) + \mathbf{i}_z (w_0 + \zeta).$$

The displacement at P as assumed in Eqs. (7.3.1) can be written in the vector form

$$\mathbf{u} = \mathbf{i}_x (u + \zeta \psi) + \mathbf{i}_y (v + \zeta \varphi) + \mathbf{i}_z w.$$

The radius vector after deformation is simply

$$\mathbf{r}' = \mathbf{r} + \mathbf{u}.$$

The nonlinear strains are determined by evaluating the difference between the squares $d\mathbf{r}' \cdot d\mathbf{r}'$ and $d\mathbf{r} \cdot d\mathbf{r}$ in a similar manner as described in Section 1.1. The calculation is tedious but straightforward. After simplification, the nonlinear strains are found to be

$$\epsilon_{xx} = \epsilon_{xx}^{(0)} + \zeta\, \epsilon_{xx}^{(1)}$$
$$= \frac{\partial u}{\partial x} + \frac{\partial w_0}{\partial x} \frac{\partial w}{\partial x} + \frac{1}{2} \left(\frac{\partial w}{\partial x} \right)^2 + \zeta\, \frac{\partial \psi}{\partial x}$$

$$\epsilon_{yy} = \epsilon_{yy}^{(0)} + \zeta\, \epsilon_{yy}^{(1)}$$
$$= \frac{\partial v}{\partial y} + \frac{\partial w_0}{\partial y} \frac{\partial w}{\partial y} + \frac{1}{2} \left(\frac{\partial w}{\partial y} \right)^2 + \zeta\, \frac{\partial \phi}{\partial y}$$

$$\epsilon_{xy} = \epsilon_{xy}^{(0)} + \zeta\, \epsilon_{xy}^{(1)}$$
$$= \frac{1}{2} \left[\left(\frac{\partial u}{\partial y} + \frac{\partial v}{\partial x} \right) + \frac{\partial w_0}{\partial y} \frac{\partial w}{\partial x} \right.$$
$$\left. + \frac{\partial w_0}{\partial x} \frac{\partial w}{\partial y} + \frac{\partial w}{\partial x} \frac{\partial w}{\partial y} + \frac{1}{2} \zeta \left(\frac{\partial \psi}{\partial y} + \frac{\partial \phi}{\partial x} \right) \right]$$

$$\epsilon_{\zeta x} = \epsilon_{\zeta x}^{(0)} = \frac{1}{2} \left(\psi + \frac{\partial w}{\partial x} \right)$$

$$\epsilon_{\zeta y} = \epsilon_{\zeta y}^{(0)} = \frac{1}{2} \left(\phi + \frac{\partial w}{\partial y} \right),$$

where $\epsilon_{xx}^{(0)}, \ldots$ and $\epsilon_{xx}^{(1)}, \ldots$ are nonlinear shell strains of the zeroth and first order, respectively. The shell stress–strain–displacement relations are then

$$N_x = C(\epsilon_{xx}^{(0)} + \nu\epsilon_{yy}^{(0)})$$

$$= C\left[\frac{\partial u}{\partial x} + \nu\frac{\partial v}{\partial y} + \frac{\partial w_0}{\partial x}\frac{\partial w}{\partial x} + \nu\frac{\partial w_0}{\partial y}\frac{\partial w}{\partial y}\right.$$

$$\left. + \frac{1}{2}\left(\frac{\partial w}{\partial x}\right)^2 + \frac{1}{2}\nu\left(\frac{\partial w}{\partial y}\right)^2\right]$$

$$N_y = C(\epsilon_{yy}^{(0)} + \nu\epsilon_{xx}^{(0)})$$

$$= C\left[\frac{\partial v}{\partial y} + \nu\frac{\partial u}{\partial x} + \frac{\partial w_0}{\partial y}\frac{\partial w}{\partial y} + \nu\frac{\partial w_0}{\partial x}\frac{\partial w}{\partial x}\right.$$

$$\left. + \frac{1}{2}\left(\frac{\partial w}{\partial y}\right)^2 + \frac{1}{2}\nu\left(\frac{\partial w}{\partial x}\right)^2\right]$$

$$N_{xy} = C(1-\nu)\epsilon_{xy}^{(0)}$$

$$= \frac{1}{2}C(1-\nu)\left[\frac{\partial u}{\partial y} + \frac{\partial v}{\partial x} + \frac{\partial w_0}{\partial y}\frac{\partial w}{\partial x}\right. \tag{7.3.4}$$

$$\left. + \frac{\partial w_0}{\partial x}\frac{\partial w}{\partial y} + \frac{\partial w}{\partial x}\frac{\partial w}{\partial y}\right]$$

$$M_x = D(\epsilon_{xx}^{(1)} + \nu\epsilon_{yy}^{(1)}) = D\left(\frac{\partial \psi}{\partial x} + \nu\frac{\partial \phi}{\partial y}\right)$$

$$M_y = D(\epsilon_{yy}^{(1)} + \nu\epsilon_{xx}^{(1)}) = D\left(\frac{\partial \phi}{\partial y} + \nu\frac{\partial \psi}{\partial x}\right)$$

$$M_{xy} = 2D(1-\nu)\epsilon_{xy}^{(1)} = D(1-\nu)\left(\frac{\partial \psi}{\partial y} + \frac{\partial \phi}{\partial x}\right)$$

$$Q_x = 4\kappa Gh\epsilon_{\zeta x}^{(0)} = 2\kappa Gh\left(\psi + \frac{\partial w}{\partial x}\right)$$

$$Q_y = 4\kappa Gh\epsilon_{\zeta y}^{(0)} = 2\kappa Gh\left(\phi + \frac{\partial w}{\partial y}\right).$$

Substitution of these relations in the stress equations of motion written from Eq. (7.3.3) readily yields the displacement equations of motion.

7.3.3 Suppression of Transverse Shear

The original Marguerre equations can be deduced from the above results as a special case by suppressing the transverse shear effect. They also can be derived directly from Eq. (1.4.8) by assuming the displacements in the form

$$u_x = u - \zeta\frac{\partial w}{\partial x}, \quad u_y = v - \zeta\frac{\partial w}{\partial y}, \quad u_z = w. \tag{7.3.5}$$

The pseudo-variational equation of motion then is reduced to the following two-dimensional form:

$$
\int_{t_0}^{t_1} dt \int_A \int \Bigg(\left\{ \frac{\partial N_x}{\partial x} + \frac{\partial N_{xy}}{\partial y} + P_x^{(0)} + f_x^{(0)} - 2\rho h \ddot{u} \right\} \delta u
$$

$$
+ \left\{ \frac{\partial N_{xy}}{\partial x} + \frac{\partial N_y}{\partial y} + P_y^{(0)} + f_y^{(0)} - 2\rho h \ddot{v} \right\} \delta v
$$

$$
+ \Bigg\{ \frac{\partial^2 M_x}{\partial x^2} + 2 \frac{\partial^2 M_{xy}}{\partial x \partial y} + \frac{\partial^2 M_y}{\partial y^2}
$$

$$
+ \frac{\partial}{\partial x} \left[N_x \frac{\partial}{\partial x}(w_0 + w) + N_{xy} \frac{\partial}{\partial y}(w_0 + w) \right]
$$

$$
+ \frac{\partial}{\partial y} \left[N_{xy} \frac{\partial}{\partial x}(w_0 + w) + N_y \frac{\partial}{\partial y}(w_0 + w) \right]
$$

$$
+ \frac{\partial P_x^{(1)}}{\partial x} + \frac{\partial P_y^{(1)}}{\partial y} + P_z^{(0)}
$$

$$
+ \frac{\partial f_x^{(1)}}{\partial x} + \frac{\partial f_y^{(1)}}{\partial y} + f_z^{(0)}
$$

$$
- 2\rho h \ddot{w} + \tfrac{2}{3}\rho h^3 \nabla^2 \ddot{w} \Bigg\} \delta w \Bigg) \, dx \, dy
$$

$$
- \int_{t_0}^{t_1} dt \oint_{C_p} \Bigg\{ \left[N_x - p_n^{(0)} \right] \delta u_n + \left[N_{ns} - p_s^{(0)} \right] \delta u_s
$$

$$
+ \left[M_n - p_n^{(1)} \right] \delta \left(\frac{\partial w}{\partial n} \right) + \left[\frac{\partial M_n}{\partial n} + 2 \frac{\partial M_{ns}}{\partial s} \right.
$$

$$
+ N_n \frac{\partial}{\partial n}(w_0 + w) + N_{ns} \frac{\partial}{\partial s}(w_0 + w)
$$

$$
+ P_n^{(1)} + f_n^{(1)} - p_z^{(0)}
$$

$$
- \frac{\partial p_s^{(1)}}{\partial s} + \frac{2}{3}\rho h^3 \frac{\partial \ddot{w}}{\partial n} \Bigg] \delta w \Bigg\} \, ds = 0,
$$

where

$$
M_x = -D\left(\frac{\partial^2 w}{\partial x^2} + \nu \frac{\partial^2 w}{\partial y^2} \right)
$$

$$
M_y = -D\left(\frac{\partial^2 w}{\partial y^2} + \nu \frac{\partial^2 w}{\partial x^2} \right)
$$

$$
M_{xy} = -D(1 - \nu)\frac{\partial^2 w}{\partial x \partial y}.
$$

N_x, N_y, N_{xy} are still given by Eqs. (7.3.4), and Q_x and Q_y no longer appear.

7.4   Remarks on the Variational Equations of Motion

Now that some applications of the pseudo-variational equation of motion (1.4.8) have been explored, a few remarks are in order. First of all, we note that while both Eqs. (1.4.7) and (1.4.8) are pseudo-variational equations, the latter has a simpler form and enjoys a further distinction in that we were able to derive from it directly the von Kármán and Marguerre equations for large deflections of plates and shallow shells. Furthermore, we were also able to extend both sets of equations by incorporating the transverse shear effect. These have provided a justification for the use of the pseudo-variational equation (1.4.8).

The remaining remarks in this section apply to both linear and nonlinear variational equations of motion. Insofar as a variational equation of motion is exactly equivalent to the original Hamilton's principle from which it was derived, it is important to realize that both the volume and surface integrals in the variational equation are parts of a total package, including even the negative sign between the integrals. In general, both integrals must be used jointly together. This has been demonstrated when the variational equation of motion is employed to derive equations of plates and shallow shells, including particularly those of the classical type in which the assumed three- dimensional displacements in terms of the plate-displacements are coupled (Yu 1995a,b).

Furthermore, with the use of variational equations of motion, derivation of two-dimensional plate and shell equations from three-dimensional elasticity has become straightforward, as demonstrated in both our early and recent (Yu 1991, 1992a,b, 1995a,b) works. This is in contrast to the use of Hamilton's principle or any other variational principle, during the process of which one often does not get to see the end of the tunnel until the entire mathematical manipulation is completed. More importantly, as we start with a variational principle, approximations are often introduced without our paying special attention, and we can be misled to believe that exact results are always obtained at the end of the tunnel. In contrast, starting with the variational equation helps eliminate such illusion since the user is forced to recognize the approximations that are introduced from the very beginning.

At this point we remark that, just as is the variational principle, the variational equation of motion is a very powerful tool for carrying out approximations. Associated with nonlinear vibration analysis of plates and shells, we have proposed an integrated procedure with the use of the variational equation of motion (Yu 1963), which consists of a sequence of variational approximations. The procedure begins with an approximation in the thickness direction and yields the plate and shell equations as demonstrated in this book. But this is only the first step in structural modeling. Additional steps may involve subsequent variational approximations with respect to the remaining space coordinates, and even with respect to time, whenever needed. In fact, a proper combination of the variational approximations is better assured with the use of the variational equation of motion, as demonstrated in Section 6.2 with the application of the Galerkin method to a linear vibration analysis.

Additional variational approximations may also be executed through the use of finite and boundary element methods or other approximate numerical methods. This is to make meaningful and effective connections between the variational principles and variational equations on the one hand and the numerical methods on the other hand. In an early work (Yu 1974), we have expressed an interest in such a connection by exploring the flexibility provided by the variational equation in a finite element analysis of nonlinear vibrations of beams with large deflections. Indeed, the increasing importance of this kind of connection was well demonstrated by Washizu's book (1968, 1975, 1982) in which coverage of the finite-element method eventually became a complete separate part in his third and last edition.

References

Chia, C.Y. (1980) *Nonlinear Analysis of Plates.* McGraw-Hill, New York.

Dym, C.L. and I.H. Shames. (1973) *Solid Mechanics: A Variational Approach.* McGraw-Hill, New York.

Eringen, A.C. (1955) On the Nonlinear Oscillations of Viscoelastic Plates. *Journal of Applied Mechanics,* Vol. 22, pp. 563–567.

Fung, Y.C. (1965) *Foundations of Solid Mechanics.* Prentice-Hall, Englewood Cliffs, New Jersey.

Langhaar, H.L. (1962) *Energy Methods in Applied Mechanics.* Wiley, New York.

Marguerre, K. (1938) Zur Theorie Der Gekrummten Platte Grosser Formanderung. In: *Proceedings of the 5th International Congress of Applied Mechanics,* pp. 93–101.

Medwadowski, S.J. (1958) A Refined Theory of Elastic, Orthotropic Plates. *Journal of Applied Mechanics,* Vol. 25, pp. 437–443.

Novozhilov, V.V. (1948) *Foundations of the Nonlinear Theory of Elasticity.* (English translation, Graylock, Rochester, New York, 1953.)

Reddy, J.N. (1984) A Refined Nonlinear Theory of Plates with Transverse Shear Deformation. *International Journal of Solids and Structures,* Vol. 20, pp. 881–896.

Timoshenko, S. (1921) On the Correction for Shear of the Differential Equation for Transverse Vibrations of Prismatic Bars. *Philosophical Magazine,* Series 6, Vol. 41, pp. 744–746.

Timoshenko, S. (1940) *Theory of Plates and Shells,* McGraw-Hill, New York.

Washizu, K. (1968) *Variational Methods in Elasticity and Plasticity,* Pergamon, New York (second edition, 1975; third edition, 1982).

von Kármán, T. (1910) Festigkeitsproblems im Maschinenbau. *Encyklopadie der mathematischen Wissenschafter*, Vol. 4, pp. 348–352.

Yu, Y.Y. (1963) Application of Variational Equation of Motion to the Nonlinear Analysis of Homogeneous and Layered Plates and Shells. *Journal of Applied Mechanics*, Vol. 30, pp. 79–86.

Yu, Y.Y. (1974) Application of Variational and Galerkin Equations to Linear and Nonlinear Finite Element Analysis. In: *Proceedings of the 25th Congress of International Astronautical Federation*, Amsterdam.

Yu, Y.Y. (1991) On Equations for Large Deflections of Elastic Plates and Shallow Shells. *Mechanics Research Communications*, Vol. 18, pp. 373–384.

Yu, Y.Y. (1992a) Equations for Large Deflections of Homogeneous and Layered Beams with Applications to Chaos and Acoustic Radiation. *Composites Engineering*, Vol. 2, pp. 117–136.

Yu, Y.Y. (1992b) Generalized Variational Equations of Motion in Nonlinear Anisotropic Elasticity and Plate Theories. In: *Collection of Papers, a Volume Dedicated to the 80th Birthday of Academician V. V. Novozhilov*, edited by N.S. Solomenko, pp. 63–74. Shipbuilding Publishing House, St. Petersburg, Russia.

Yu, Y.Y. (1995a) On the Ordinary, Generalized, and Pseudo-Variational Equations of Motion in Nonlinear Elasticity, Piezoelectricity, and Classical Plate Theories. *Journal of Applied Mechanics*, Vol. 62, pp. 471–478.

Yu, Y.Y. (1995b) Some Recent Advances in Linear and Nonlinear Dynamical Modeling of Elastic and Piezoelectric Plates. *Journal of Intelligent Material Systems and Structures*, Vol. 6, pp. 237–254.

8

Nonlinear Modeling and Vibrations of Sandwiches and Laminated Composites

In the preceding chapter, a method of developing nonlinear dynamical models has been applied to homogeneous beams, plates, and shallow shells through the use of a pseudo-variational equation of motion in nonlinear elasticity. The famous Timoshenko beam equations, von Kármán plate equations, and Marguerre shallow shell equations have been rederived and extended. In this chapter, the method is further applied to layered plates, including sandwiches and laminated composites. The equations derived are next applied to the analysis of nonlinear vibrations.

Our early efforts dealt with both linear and nonlinear vibrations of homogeneous and sandwich plates and shells (Yu 1962, 1963). Later, nonlinear oscillations of laminated composite plates were investigated by Wu and Vinson (1969), forced nonlinear vibrations of damped sandwich beams by Kovac et al. (1971), and nonlinear vibrations of three-layered beams with viscoelastic cores by Hyer et al. (1976, 1978). As cited earlier in Chapters 4 and 5, the recent review (Yu 1989, 1992a) has covered not only our early results on linear and nonlinear vibrations of sandwiches, but also the impressive progress that was made in laminated composites by other previous workers. More recently, we have constructed nonlinear dynamical models for large deflections of various types of beams and plates (Yu 1991, 1992b,c,d, 1993, 1995a,b), including laminated composite beams and plates, and sandwich beams with either isotropic or orthotropic core and with either isotropic or laminated composite facings. Applications to the analysis of vibrations, chaos, and acoustic radiation have also been explored.

Thus, refined equations for large deflections of a sandwich plate are developed in Section 8.1 and applied to nonlinear vibrations analysis in Section 8.2. Similarly, classical and refined equations for large deflections of a laminated composite plate are derived in Section 8.3 and applied to the nonlinear vibration analysis of an orthotropic symmetric laminate in Section 8.4. Finally, to demonstrate the combined use of sandwich and laminated composites, we derive in the last section of this chapter the refined equations for large deflections of a sandwich beam with laminated composite facings and an orthotropic core.

8.1   Equations for Large Deflections of a Sandwich Plate

Linear equations and vibrations of a sandwich plate have been treated in detail in Chapter 4. Nonlinear equations for large deflections of a sandwich plate were derived in our early work (Yu 1962) and are rederived here more directly from the new pseudo-variational equation (1.4.8). Since it is essential to include the transverse shear effect in the sandwich core, the nonlinear equations obtained are a generalization of the simplified refined linear equations for an isotropic sandwich plate with membrane facings.

8.1.1   Variational Equation of Motion

The displacements in the three layers of a sandwich plate are taken in the form

$$u_{x1} = u + z\psi, \qquad u_{x2}, u_{x3} = u \mp h_1\psi$$

$$u_{y1} = v + z\varphi, \qquad u_{y2}, u_{y3} = v \mp h_1\varphi \tag{8.1.1}$$

$$u_{z1} = u_{z2} = u_{z3} = w$$

and the rotations in the form

$$\omega_{zx1} = \omega_{zx2} = \omega_{zx3} = -\frac{\partial w}{\partial x}$$

$$\omega_{zy1} = \omega_{zy2} = \omega_{zy3} = \frac{\partial w}{\partial y}. \tag{8.1.2}$$

Equations (8.1.1) were used in our early work (Yu 1962), although an alternative set of displacements was also introduced. The rotations in Eqs. (8.1.2) differ somewhat from those used in the early work.

Substituting Eqs. (8.1.1) and (8.1.2) into the new pseudo-variational equation (1.4.8) and carrying out integration with respect to z over the total thickness of the sandwich, we find

$$\int_{t_0}^{t_1} dt \int \int_A \left(\left\{ \frac{\partial}{\partial x}(N_{x1} + N_{x2} + N_{x3}) + \frac{\partial}{\partial y}(N_{yx1} + N_{yx2} + N_{yx3}) \right.\right.$$

$$\left. + P_x^{(0)} + f_x^{(0)} - 2(\rho_1 h_1 + \rho_2 h_2)\ddot{u} \right\} \delta u$$

$$+ \left\{ \frac{\partial}{\partial x}(N_{xy1} + N_{xy2} + N_{xy3}) + \frac{\partial}{\partial y}(N_{y1} + N_{y2} + N_{y3}) \right.$$

$$\left. + P_y^{(0)} + f_y^{(0)} - 2(\rho_1 h_1 + \rho_2 h_2)\ddot{v} \right\} \delta v$$

$$
+\left\{\frac{\partial}{\partial x}\left[M_{x1} + h_1(N_{x3} - N_{x2})\right] + \frac{\partial}{\partial y}\left[M_{yx1} + h_1(N_{yx3} - N_{yx2})\right]\right.
$$

$$
\left. - Q_{x1} + P_x^{(1)} + f_x^{(1)} - \left(\frac{2}{3}\rho_1 h_1^3 + 2\rho_2 h_2 h_1^2\right)\ddot{\psi}\right\}\delta\psi
$$

$$
+\left\{\frac{\partial}{\partial x}\left[M_{xy1} + h_1(N_{xy3} - N_{xy2})\right] + \frac{\partial}{\partial y}\left[M_{y1} + h_1(N_{y3} - N_{y2})\right]\right.
$$

$$
\left. - Q_{y1} + P_y^{(1)} + f_y^{(1)} - \left(\frac{2}{3}\rho_1 h_1^3 + 2\rho_2 h_2 h_1^2\right)\ddot{\phi}\right\}\delta\phi
$$

$$
+\left\{\frac{\partial}{\partial x}\left[Q_{x1} + (N_{x1} + N_{x2} + N_{x3})\frac{\partial w}{\partial x} + (N_{xy1} + N_{xy2} + N_{xy3})\frac{\partial w}{\partial y}\right]\right.
$$

$$
+ \frac{\partial}{\partial y}\left[Q_{y1} + (N_{yx1} + N_{yx2} + N_{yx3})\frac{\partial w}{\partial x} + (N_{y1} + N_{y2} + N_{y3})\frac{\partial w}{\partial y}\right]
$$

$$
\left.\left. + P_z^{(0)} + f_z^{(0)} - 2(\rho_1 h_1 + \rho_2 h_2)\ddot{w}\right\}\delta w\right)dx\,dy
$$

$$
-\int_{t_0}^{t_1} dt \oint_{C_p}\left\{[N_{n1} + N_{n2} + N_{n3} - p_n^{(0)}]\delta u_n\right.
$$

$$
+ [N_{ns1} + N_{ns2} + N_{ns3} - p_s^{(0)}]\delta u_s
$$

$$
+ [M_{n1} + h_1(N_{n3} + N_{n2}) - p_{n1}^{(1)} + h_1(p_{n3}^{(0)} - p_{n2}^{(0)})]\delta\psi_n
$$

$$
+ [M_{ns1} + h_1(N_{ns3} + N_{ns2}) - p_{s1}^{(1)} - h_1(p_{s3}^{(0)} - p_{s2}^{(0)})]\delta\psi_s
$$

$$
+ [Q_{n1} + (N_{n1} + N_{n2} + N_{n3})\frac{\partial w}{\partial n}
$$

$$
\left. + (N_{ns1} + N_{ns2} + N_{ns3})\frac{\partial w}{\partial s} - p_z^{(0)}]\delta w\right\}ds = 0, \qquad (8.1.3)
$$

where the plate stresses are defined in the same way as in the linear case. Equation (8.1.3) is the pseudo-variational equation of motion for large deflections of a sandwich plate from which the stress equations of motion and traction boundary conditions may be written in the usual manner.

8.1.2 Displacements, Strains, and Stresses

For the various layers in the sandwich, the three-dimensional nonlinear stress–strain relations are

$$
\sigma_{xxi} = \frac{E_i}{1 - v_i^2}(\epsilon_{xxi} + v_i\epsilon_{yyi})
$$

$$
\sigma_{yyi} = \frac{E_i}{1 - v_i^2}(\epsilon_{yyi} + v_i\epsilon_{xxi})
$$

$$
(i = 1, 2, 3) \qquad (8.1.4)
$$

$$
\sigma_{xyi} = 2\mu_i\epsilon_{xyi}
$$

$$
\sigma_{xz1} = 2\mu_{z1}\epsilon_{xz1}
$$

$$
\sigma_{yz1} = 2\mu_{z1}\epsilon_{yz1},
$$

where μ_i is the shear modulus in the xy-plane, μ_{z1} is that of the core in the transverse direction, and other elastic constants are the same as usual. The nonlinear strain–displacement relations are

$$\epsilon_{xxi} = \frac{\partial u_{xi}}{\partial x} + \frac{1}{2}\omega_{xzi}^2$$

$$\epsilon_{yyi} = \frac{\partial u_{yi}}{\partial y} + \frac{1}{2}\omega_{yzi}^2$$

$$\epsilon_{xyi} = \frac{1}{2}\left(\frac{\partial u_{yi}}{\partial x} + \frac{\partial u_{xi}}{\partial y} - \omega_{xzi}\omega_{yzi}\right) \qquad (i = 1, 2, 3) \qquad (8.1.5)$$

$$\epsilon_{xz1} = \frac{1}{2}\left(\frac{\partial u_{x1}}{\partial z} + \frac{\partial u_{z1}}{\partial x}\right)$$

$$\epsilon_{yz1} = \frac{1}{2}\left(\frac{\partial u_{y1}}{\partial z} + \frac{\partial u_{z1}}{\partial y}\right).$$

Substitution of Eqs. (8.1.5) into (8.1.4) and integration of the result according to the definitions of the plate stresses yield

$$N_{x1} = C_1\left[\frac{\partial u}{\partial x} + \nu_1\frac{\partial v}{\partial y} + \frac{1}{2}\left(\frac{\partial w}{\partial x}\right)^2 + \frac{1}{2}\nu_1\left(\frac{\partial w}{\partial y}\right)^2\right]$$

$$N_{y1} = C_1\left[\frac{\partial v}{\partial y} + \nu_1\frac{\partial u}{\partial x} + \frac{1}{2}\left(\frac{\partial w}{\partial y}\right)^2 + \frac{1}{2}\nu_1\left(\frac{\partial w}{\partial x}\right)^2\right]$$

$$N_{xy1} = \frac{1}{2}C_1(1 - \nu_1)\left[\frac{\partial v}{\partial x} + \frac{\partial u}{\partial y} + \frac{\partial w}{\partial x}\frac{\partial w}{\partial y}\right]$$

$$N_{x2}, N_{x3} = C_2\left[\frac{\partial u}{\partial x} + \nu_2\frac{\partial v}{\partial y} + \frac{1}{2}\left(\frac{\partial w}{\partial x}\right)^2 + \frac{1}{2}\nu_2\left(\frac{\partial w}{\partial y}\right)^2\right.$$

$$\left.\mp h_1\left(\frac{\partial \psi}{\partial x} + \nu_2\frac{\partial \phi}{\partial y}\right)\right]$$

$$N_{y2}, N_{y3} = C_2\left[\frac{\partial v}{\partial y} + \nu_2\frac{\partial u}{\partial x} + \frac{1}{2}\left(\frac{\partial w}{\partial y}\right)^2 + \frac{1}{2}\nu_2\left(\frac{\partial w}{\partial x}\right)^2\right.$$

$$\left.\mp h_1\left(\frac{\partial \phi}{\partial y} + \nu_2\frac{\partial \psi}{\partial x}\right)\right] \qquad (8.1.6)$$

$$N_{xy2}, N_{xy3} = \frac{1}{2}C_2(1 - \nu_2)\left[\frac{\partial v}{\partial x} + \frac{\partial u}{\partial y} + \frac{\partial w}{\partial x}\frac{\partial w}{\partial y}\right.$$

$$\left.\mp h_1\left(\frac{\partial \phi}{\partial x} + \frac{\partial \psi}{\partial y}\right)\right]$$

$$M_{x1} = D_1\left(\frac{\partial \psi}{\partial x} + \nu_1\frac{\partial \phi}{\partial y}\right)$$

$$M_{y1} = D_1 \left(\frac{\partial \phi}{\partial y} + \nu_1 \frac{\partial \psi}{\partial x} \right)$$

$$M_{xy1} = D_1 (1 - \nu_1) \left(\frac{\partial \phi}{\partial x} + \frac{\partial \psi}{\partial y} \right)$$

$$Q_{x1} = 2\kappa_1 \mu_{z1} h_1 \left(\psi + \frac{\partial w}{\partial x} \right)$$

$$Q_{y1} = 2\kappa_1 \mu_{z1} h_1 \left(\phi + \frac{\partial w}{\partial y} \right),$$

where

$$C_1 = \frac{2E_1 h_1}{1 - \nu_1^2}$$

$$C_2 = \frac{E_2 h_2}{1 - \nu_2^2}$$

$$D_1 = \frac{2E_1 h_1^3}{3(1 - \nu_1^2)}.$$

These are the plate stress–displacement relations by means of which the stress equations of motion of the sandwich plate can be transformed readily into the displacement equations of motion. The latter will be applied to the nonlinear vibration analysis of a sandwich plate in the next section.

8.2   Nonlinear Vibration of a Sandwich Plate

On the basis of the refined equations derived in the preceding section, the nonlinear free vibration of a sandwich plate in plane strain and with immovable hinged edges was investigated in our early work (Yu 1962) and will be covered in this section. In particular, the effect of transverse shear will be assessed. The nonlinear vibration of a rectangular sandwich plate with four immovable hinged edges was also treated in the early work but will not be covered here. Results for the sandwich plate reduce readily to those for a homogeneous plate as a special case.

8.2.1   Nonlinear Free Vibration

For the plane strain case with a functional dependence on x, the following displacement equations of motion are obtained from Eqs. (8.1.3) and (8.1.6):

$$\frac{\partial^2 u}{\partial x^2} + \frac{\partial w}{\partial x} \frac{\partial^2 w}{\partial x^2} = 0 \tag{8.2.1}$$

$$C_2 h_1^2 \frac{\partial^2 \psi}{\partial x^2} - \kappa_1 \mu_{z1} h_1 \left(\psi + \frac{\partial w}{\partial x} \right) = 0 \tag{8.2.2}$$

$$\kappa_1 \mu_{z1} h_1 \left(\frac{\partial \psi}{\partial x} + \frac{\partial^2 w}{\partial x^2} \right) + C_2 \left[\frac{\partial u}{\partial x} + \frac{1}{2} \left(\frac{\partial w}{\partial x} \right)^2 \right] \frac{\partial^2 w}{\partial x^2}$$

$$- (\rho_1 h_1 + \rho_2 h_2) \ddot{w} = 0 \qquad (8.2.3)$$

For nonlinear vibration, the displacements are assumed to be in the form

$$w = W \tau(t) \sin \frac{\lambda x}{h}, \qquad \psi = \Psi \tau(t) \cos \frac{\lambda x}{h}, \qquad (8.2.4)$$

where $\lambda = n\pi h/\ell$, with $n = 1, 2, 3, \ldots$ designating the number of half-waves in the span length ℓ of the plate. These satisfy the conditions of 0 deflection and 0 moment at the hinged edges and are similar to those assumed for the linear case, the only difference being that the function of time $\tau(t)$ is now not known. Substitution of w from Eqs. (8.2.4) into Eq. (8.2.1) and subsequent integration yield

$$u = -W^2 \tau^2 \frac{\lambda}{8h} \sin \frac{2\lambda x}{h}, \qquad (8.2.5)$$

where the integration constants have been taken equal to 0 to satisfy the boundary conditions $u = 0$ at $x = 0, \ell$ for the immovable edges. Next, we substitute Eqs. (8.2.4) into (8.2.2) and find the following relation between the amplitudes Ψ and W:

$$\Psi = -\frac{\lambda W}{h} \frac{1}{1 + r_2 r_h \lambda^2 / \kappa_1 (1 + r_h)^2} \qquad (8.2.6)$$

where, as before, the ratios are $r_2 = C_2/\mu_{z1} h_2$ and $r_h = h_2/h_1$. Finally, by virtue of Eqs. (8.2.4) through (8.2.6), Eq. (8.2.3) yields, with the ratio $r_\rho = \rho_2/\rho_1$,

$$\ddot{\tau} + \frac{\mu_{z1}}{\rho_1 h^2} \frac{r_2 r_h \lambda^4}{1 + r_\rho r_h} \left[\frac{1}{(1 + r_h)^2 + r_2 r_h \lambda^2 / \kappa_1} \tau + \left(\frac{W}{2h} \right)^2 \tau^3 \right] = 0 \qquad (8.2.7)$$

This is the governing equation for the time function $\tau(t)$. The nonlinear term contains the square of the ratio $W/2h$ as a factor.

Equation (8.2.7) has the standard nonlinear form

$$\ddot{\tau} + \alpha \tau + \beta \tau^3 = 0. \qquad (8.2.8)$$

The exact solution of this equation is well known, according to which the ratio between the nonlinear frequency ω and the associated linear frequency $\omega_0 = \sqrt{a}$ is given by

$$\frac{\omega}{\omega_0} = \frac{\pi}{2} \frac{\sqrt{1 + \beta/\alpha}}{K(k, \pi/2)},$$

where K is the complete elliptic integral of the first kind, with a parameter

$$k = \frac{1}{\sqrt{2(1 + \alpha/\beta)}}.$$

A simple approximate solution of Eq. (8.2.8) is further obtainable by the use of Duffing's method, which is so popular that Eq. (8.2.8) often is referred to as a *Duffing's equation*. According to the method, we assume

$$\tau = \cos \omega t$$

so that

$$\tau^3 = \cos^3 \omega t = 3 \cos \frac{\omega t}{4} + \cos \frac{3\omega t}{4}.$$

When τ and τ^3 are substituted into Eq. (8.2.7) or (8.2.8) and the coefficient of $\cos \omega t$ is equated to zero, we find

$$\frac{\omega}{\omega_0} = \sqrt{1 + \frac{3\beta}{4\alpha}} = \sqrt{1 + \frac{3}{4}\left(\frac{W}{2h}\right)^2 \left\{(1 + r_h)^2 + \frac{r_2 r_h \lambda^2}{\kappa_1}\right\}}. \qquad (8.2.9)$$

The nonlinear frequency thus increases with the square of the amplitude W and also with r_2, r_h, and λ.

8.2.2   Effect of Transverse Shear

To determine the effect of transverse shear on the nonlinear frequency, we rewrite Eq. (8.2.9) in the form

$$\omega = \sqrt{\alpha + \frac{3\beta}{4}}$$

$$= \sqrt{\frac{\mu_{z1}}{\rho_1 h^2} \frac{r_2 r_h \lambda^4}{1 + r_\rho r_h} \left[\frac{1}{(1 + r_h)^2 + r_2 r_h \lambda^2 \kappa_1} + \frac{3}{4}\left(\frac{W}{2h}\right)^2\right]}.$$

$$(8.2.10)$$

When the transverse shear effect is neglected, by letting $\kappa_1 = \infty$, the result deduced from Eq. (8.2.10) is denoted by ω', and we find the ratio

$$\frac{\omega}{\omega'} = \sqrt{\frac{1}{(1 + r_h)^2 + r_2 r_h \lambda^2 \kappa_1} + \frac{3}{4}\left(\frac{W}{2h}\right)^2} \div \sqrt{\frac{1}{(1 + r_h)^2} + \frac{3}{4}\left(\frac{W}{2h}\right)^2}.$$

$$(8.2.11)$$

Since this is always smaller than 1, the shear effect is to lower the nonlinear frequency. As in the linear case, the result also indicates that the shear effect becomes negligible if

$$\frac{r_2 r_h \lambda^2}{(1 + r_h)^2} \ll 1. \qquad (8.2.12)$$

However, the error caused by neglecting the shear effect is always smaller in the nonlinear than in the linear case. This could have been expected as the shear effect is more directly associated with flexure than with extension, and the latter gains importance as the deflection increases. In fact, Eq. (8.2.11) shows that the error caused by neglecting the shear effect would be negligible if

$$\frac{3}{4}\left(\frac{W}{2h}\right)^2 \gg \frac{1}{(1+r_h)^2}. \tag{8.2.13}$$

Actually, deflections of such magnitude could not take place during the vibration of ordinary sandwich plates for which r_h is small because the strains corresponding to such deflections would be enormously large. In real cases in which the deflection is of a much smaller magnitude, the shear effect on either linear or nonlinear frequency should in general be considered.

Since r_2 and λ^2 appear together in Eqs. (8.2.9) and (8.2.11), the product $r_2\lambda^2$ may be considered as a single parameter. As $r_2\lambda^2$ increases, both ω/ω_0 and ω/ω' deviate further from unity, which indicates an increasing importance of both shear and nonlinear effects. Furthermore, since λ varies with n as well as with h/ℓ for a sandwich plate of given materials and dimensions, we have not just one value but a sequence of values of $r_2\lambda^2$, and thus also a sequence of values for each of ω/ω_0 and ω/ω' when the vibration amplitude is further given. Indeed, $r_2\lambda^2$ increases rapidly as n is assigned the values 1, 2, 3, . . . consecutively, and so does the importance of the nonlinear and shear effects.

8.2.3   Additional Results

In the preceding section, we have mentioned that a different set of displacements and rotations also was adopted in our early study (Yu 1962). Instead of the displacements in Eqs. (8.1.1), those in Eqs (4.2.6) were used. These led to a different system of nonlinear equations. When the equations were applied to the vibration analysis of a sandwich plate in plane strain, it is interesting to note that the results differed only slightly from those in Eqs. (8.2.6), (8.2.7), and (8.2.9) through (8.2.13) and were obtainable from the latter by simply replacing $(1 + r_h)^2$ by $(1 + r_h)$.

8.3   Equations for Large Deflections of a Laminated Composite Plate

Linear equations and vibrations of a laminated composite plate were discussed in Chapter 5. We now extend the discussion to large deflections of a laminate (Yu 1993).

8.3.1   Classical Equations for Large Deflections of a Laminate

The procedure of treating the laminate is the same as before except that it is now based on the nonlinear pseudo-variational equation (1.4.8) instead of the linearized version. The displacements in the laminate are assumed to be in the classical form

$$u_x = u - z\frac{\partial w}{\partial x}, \qquad u_y = v - z\frac{\partial w}{\partial y}, \qquad u_z = w, \tag{8.3.1}$$

which now reflect the coupling between extension and flexure because of large deflections. Of the three rotations, ω_{xy} already has been neglected in the pseudo-variational equation, and the other two are taken in the same form as before:

$$\omega_{zx} = -\frac{\partial w}{\partial x}, \qquad \omega_{yz} = \frac{\partial w}{\partial y}. \tag{8.3.2}$$

Substituting these displacements and rotations into Eq. (1.4.8) and carrying out integration with respect to z over the entire thickness of the laminate, we find

$$\int_{t_0}^{t_1} dt \int\int_A \left\{ \left[\frac{\partial N_x}{\partial x} + \frac{\partial N_{yx}}{\partial y} + P_x^{(0)} + f_x^{(0)} - \rho^{(0)}\ddot{u} + \rho^{(1)}\frac{\partial \ddot{w}}{\partial x} \right] \delta u \right.$$

$$+ \left[\frac{\partial N_{xy}}{\partial x} + \frac{\partial N_y}{\partial y} + P_y^{(0)} + f_y^{(0)} - \rho^{(0)}\ddot{v} + \rho^{(1)}\frac{\partial \ddot{w}}{\partial y} \right] \delta v$$

$$+ \left[\frac{\partial^2 M_x}{\partial x^2} + 2\frac{\partial^2 M_{xy}}{\partial x \partial y} + \frac{\partial^2 M_y}{\partial y^2} \right.$$

$$+ \frac{\partial}{\partial x}\left(N_x \frac{\partial w}{\partial x} + N_{xy}\frac{\partial w}{\partial y} \right) + \frac{\partial}{\partial y}\left(N_{xy}\frac{\partial w}{\partial x} + N_y\frac{\partial w}{\partial y} \right)$$

$$+ \frac{\partial P_x^{(1)}}{\partial x} + \frac{\partial P_y^{(1)}}{\partial y} + P_z^{(0)}$$

$$+ \frac{\partial f_x^{(1)}}{\partial x} + \frac{\partial f_y^{(1)}}{\partial y} + f_z^{(0)}$$

$$\left. \left. - \rho^{(0)}\ddot{w} - \rho^{(1)}\left(\frac{\partial \ddot{u}}{\partial x} - \frac{\partial \ddot{v}}{\partial y} \right) + \rho^{(2)}\nabla^2\ddot{w} \right] \delta w \right\} dx\, dy$$

$$\int_{t_0}^{t_1} dt \oint_{C_p} \left\{ [N_n - p_n^{(0)}]\delta u_n + [N_{ns} - p_s^{(0)}]\delta u_s \right.$$

$$- [M_n - p_n^{(1)}]\delta\left(\frac{\partial w}{\partial n} \right)$$

$$+ \left[\frac{\partial M_n}{\partial n} + 2\frac{\partial M_{ns}}{\partial s} + N_n\frac{\partial w}{\partial n} + N_{ns}\frac{\partial w}{\partial s} \right.$$

$$+ P_n^{(1)} + f_n^{(1)} - p_z^{(0)} - \frac{\partial p_s^{(1)}}{\partial s}$$

$$\left. \left. - \rho^{(1)}\ddot{u}_n + \rho^{(2)}\left(\frac{\partial \ddot{w}}{\partial n} \right) \right] \delta w \right\} ds = 0, \tag{8.3.3}$$

where the notations are similar to those used in the linear case:

$$[N_x, N_y, N_{xy}, M_x, M_y, M_{xy}] =$$

$$\sum_{k=1}^{n} \int_{h_{k-1}}^{h_k} [\sigma_{xx(k)}, \sigma_{yy(k)}, \sigma_{xy(k)}, \sigma_{xx(k)}z, \sigma_{yy(k)}z, \sigma_{xy(k)}z] \, dz$$

$$
\begin{aligned}
P_x^{(0)} &= p_x^+ + p_x^-, & P_y^{(0)} &= \dots, & P_z^{(0)} &= \dots, \\
P_n^{(0)} &= \dots, & P_s^{(0)} &= \dots \\
P_x^{(1)} &= h(p_x^+ - p_x^-), & P_y^{(1)} &= \dots, & & \text{(8.3.4)} \\
P_n^{(1)} &= \dots, & P_s^{(1)} &= \dots
\end{aligned}
$$

$$\left[f_x^{(0)}, f_x^{(1)}, f_y^{(0)}, f_y^{(1)}, f_z^{(0)} \right] = \sum_{k=1}^{n} \int_{h_{k-1}}^{h_k} \left[f_{x(k)}, f_{x(k)}z, f_{y(k)}, f_{y(k)}z, f_{z(k)} \right] \, dz$$

$$\left[\rho^{(0)}, \rho^{(1)}, \rho^{(2)} \right] = \sum_{k=1}^{n} \int_{h_{k-1}}^{h_k} \rho_{(k)}[1, z, z^2] \, dz$$

$$\left[p_x^{(0)}, p_x^{(1)} \right] = \sum_{k=1}^{n} \int_{h_{k-1}}^{h_k} p_x[1, z] \, dz, \dots .$$

Equation (8.3.3) is the pseudo-variational equation of motion for large deflections of a laminate, from which the stress equations of motion and traction boundary conditions can be written in the usual manner. As in the case of a linear classical plate, the transverse shear forces Q_x and Q_y have disappeared from the final result. Similarly, as in the case of a linear composite plate, $p_x^+, p_x^-, \dots$ still represent tractions at the top and bottom boundary planes of the composite plate, and tractions at the interfaces between adjacent plies have canceled out because they are actions and reactions. Also as before, the coupling inertia $\rho^{(1)}$ may not vanish for an anisotropic plate in either the linear or nonlinear case. This provides dynamic as well as static coupling between extension and flexure.

As pointed out before, the general stress–strain relations in three-dimensional nonlinear elasticity can be written in a form similar to the usual linear stress–strain relations by replacing the linear stresses and strains with components of Kirchhoff's stress tensor and Green's nonlinear strain tensor, respectively. Thus, for the state of plane stress of the kth ply in a unidirectional composite laminate, the following *nonlinear* stress–strain relations are written by generalizing the familiar linear relations:

$$
\begin{bmatrix} \sigma_{xx(k)} \\ \sigma_{yy(k)} \\ \sigma_{xy(k)} \end{bmatrix} =
\begin{bmatrix} Q_{11(k)} & Q_{12(k)} & Q_{16(k)} \\ Q_{21(k)} & Q_{22(k)} & Q_{26(k)} \\ Q_{16(k)} & Q_{26(k)} & Q_{66(k)} \end{bmatrix}
\begin{bmatrix} \epsilon_{xx} \\ \epsilon_{yy} \\ 2\epsilon_{xy} \end{bmatrix}, \qquad \text{(8.3.5)}
$$

where a factor of 2 has been attached to the nonlinear tensorial shearing strain ϵ_{xy}, and $Q_{ij(k)}$ are the off-axis stiffnesses. Corresponding to the displacements and rotations in Eqs. (8.3.1) and (8.3.2), the nonlinear strains have the typical form

$$\epsilon_{xx} = \epsilon_{xx}^{(0)} + z\,\epsilon_{xx}^{(1)},$$

where the superscripts 0 and 1 within parentheses denote strains of the zeroth and first order, respectively. For a multidirectional laminate, the resultant forces and moments are then obtained by integrating the stresses in Eqs. (8.3.5) over the thickness of the laminate according to the first of Eqs. (8.3.4). Thus, in the nonlinear case, the membrane forces and bending moments take the general form

$$
\begin{bmatrix} N_x \\ N_y \\ N_{xy} \\ M_x \\ M_y \\ M_{xy} \end{bmatrix}
=
\begin{bmatrix}
A_{11} & A_{12} & A_{16} & B_{11} & B_{12} & B_{16} \\
A_{12} & A_{22} & A_{26} & B_{12} & B_{22} & B_{26} \\
A_{16} & A_{26} & A_{66} & B_{16} & B_{26} & B_{66} \\
B_{11} & B_{12} & B_{16} & D_{11} & D_{12} & D_{16} \\
B_{12} & B_{22} & B_{26} & D_{12} & D_{22} & D_{26} \\
B_{16} & B_{26} & B_{66} & D_{16} & D_{26} & D_{66}
\end{bmatrix}
\begin{bmatrix} \epsilon_{xx}^{(0)} \\ \epsilon_{yy}^{(0)} \\ 2\epsilon_{xy}^{(0)} \\ \epsilon_{xx}^{(1)} \\ \epsilon_{yy}^{(1)} \\ 2\epsilon_{xy}^{(1)} \end{bmatrix},
\qquad (8.3.6)
$$

where A_{ij}, B_{ij}, and D_{ij} are the usual extensional stiffnesses, coupling stiffnesses, and flexural stiffnesses, respectively. Since the total number of plies in the laminate is n and the kth ply extends from $z = h_{k-1}$ to $z = h_k$, these membrane forces and bending moments are given by the typical form

$$
[N_x, \ M_x] = \sum_{k=1}^{n} \int_{h_{k-1}}^{h_k} [\sigma_{xx(k)}, \ \sigma_{xx(k)}z]\, dz,
$$

and the stiffnesses are given by

$$
[A_{ij}, \ B_{ij}, \ D_{ij}] = \sum_{k=1}^{n} Q_{ij(k)} \left[(h_k - h_{k-1}), \ \frac{(h_k^2 - h_{k-1}^2)}{2}, \ \frac{(h_k^3 - h_{k-1}^3)}{3} \right].
$$

In the simplified nonlinear case of small strains and large rotations, the three-dimensional strain– displacement relations needed here have the form

$$
\epsilon_{xx} = \frac{\partial u_x}{\partial x} + \frac{1}{2}\omega_{zx}^2
$$

$$
\epsilon_{yy} = \frac{\partial u_y}{\partial y} + \frac{1}{2}\omega_{yz}^2
\qquad (8.3.7)
$$

$$
\epsilon_{xy} = \frac{1}{2}\left(\frac{\partial u_x}{\partial y} + \frac{\partial u_y}{\partial x} - \omega_{zx}\omega_{yz} \right).
$$

By virtue of Eqs. (8.3.1) and (8.3.2), these become

$$
\epsilon_{xx} = \frac{\partial u}{\partial x} - z\frac{\partial^2 w}{\partial x^2} + \frac{1}{2}\left(\frac{\partial w}{\partial x} \right)^2, \qquad \epsilon_{yy} = \ldots
$$

$$
\epsilon_{xy} = \frac{1}{2}\left(\frac{\partial u}{\partial y} + \frac{\partial v}{\partial x} \right) - z\frac{\partial^2 w}{\partial x \partial y} + \frac{1}{2}\frac{\partial w}{\partial x}\frac{\partial w}{\partial y},
$$

$$
(8.3.8)
$$

from which the nonlinear plate strains are

$$\epsilon_{xx}^{(0)} = \frac{\partial u}{\partial x} + \frac{1}{2}\left(\frac{\partial w}{\partial x}\right)^2$$

$$\epsilon_{yy}^{(0)} = \dots$$

$$\epsilon_{xy}^{(0)} = \frac{1}{2}\left(\frac{\partial u}{\partial y} + \frac{\partial v}{\partial x}\right) + \frac{1}{2}\frac{\partial w}{\partial x}\frac{\partial w}{\partial y}$$

$$\epsilon_{xx}^{(1)} = -\frac{\partial^2 w}{\partial x^2}$$

$$\epsilon_{yy}^{(1)} = \dots$$

$$\epsilon_{xy}^{(1)} = -\frac{\partial^2 w}{\partial x \partial y}.$$

$$(8.3.9)$$

Substitution of Eqs. (8.3.9) into (8.3.6) yields the plate stress–displacement relations, by means of which the stress equations of motion are transformed readily into the displacement equations of motion for the laminate. However, the latter will not be written out here.

8.3.2 Reduction to a Single-Layered Plate

It is interesting to note that the variational equation (8.3.3) for a multiple-layered laminate can be applied to a single-layered plate by simply letting

$$\rho^{(0)} = 2\rho h, \qquad \rho^{(1)} = 0, \qquad \rho^{(2)} = \frac{2}{3}\rho h^3.$$

The stress equations of motion written from Eq. (8.3.3) for this reduced special case correspond to the von Kármán equations for large deflections of a homogeneous plate.

8.3.3 Refined Equations for Large Deflections of a Laminate

To derive the refined equations for large deflections of the same laminated composite plate, the transverse shear effect is included by taking the three-dimensional displacements in the form

$$u_x = u + z\psi, \qquad u_y = v + z\varphi, \qquad u_z = w, \qquad (8.3.10)$$

where u, v, w, ψ, and ϕ are the two-dimensional plate displacements and are independent of z. The rotations still are taken in the same form as in Eqs. (8.3.2).

By substituting Eqs. (8.3.10) and (8.3.2) into the pseudo-variational equation (1.4.8) and carrying out integration with respect to z over the entire thickness of the laminate, there results

$$\int_{t_0}^{t_1} dt \int\int_A \left\{ \left[\frac{\partial N_x}{\partial x} + \frac{\partial N_{xy}}{\partial y} + P_x^{(0)} + f_x^{(0)} - \rho^{(0)}\ddot{u} - \rho^{(1)}\ddot{\psi} \right] \delta u \right.$$

$$+ \left[\frac{\partial N_{xy}}{\partial x} + \frac{\partial N_y}{\partial y} + P_x^{(0)} + f_y^{(0)} - \rho^{(0)}\ddot{v} - \rho^{(1)}\ddot{\phi} \right] \delta v$$

$$+ \left[\frac{\partial M_x}{\partial x} + \frac{\partial M_{xy}}{\partial y} - Q_x + P_x^{(1)} + f_x^{(1)} - \rho^{(1)}\ddot{u} - \rho^{(2)}\ddot{\psi} \right] \delta \psi$$

$$+ \left[\frac{\partial M_{xy}}{\partial x} + \frac{\partial M_y}{\partial y} - Q_y + P_y^{(1)} + f_y^{(1)} - \rho^{(1)}\ddot{v} - \rho^{(2)}\ddot{\phi} \right] \delta \phi$$

$$+ \left[\frac{\partial}{\partial x}\left(Q_x + N_x \frac{\partial w}{\partial x} + N_{xy}\frac{\partial w}{\partial y} \right) \right.$$

$$+ \frac{\partial}{\partial y}\left(Q_y + N_{xy}\frac{\partial w}{\partial x} + N_y \frac{\partial w}{\partial y} \right) \tag{8.3.11}$$

$$\left. \left. + P_z^{(0)} + f_z^{(0)} - 2\rho h \ddot{w} \right] \delta w \right\} dx\, dy$$

$$- \int_{t_0}^{t_1} dt \oint_{C_p} \left\{ [N_n - p_n^{(0)}]\delta u_n + [N_{ns} - p_s^{(0)}]\delta u_s \right.$$

$$+ [M_n - p_n^{(1)}]\delta \psi_n + [M_{ns} - p_s^{(1)}]\delta \psi_s$$

$$\left. + \left[Q_n + N_n \frac{\partial w}{\partial n} + N_{ns}\frac{\partial w}{\partial s} - p_z^{(0)} \right] \delta w \right\} ds = 0,$$

where C_p is that part of the contour C on which tractions are prescribed, and most of the other notations are the same as before. It is noted that the transverse shearing forces have made their appearance in Eq. (8.3.11).

Equation (8.3.11) is the pseudo-variational equation of motion for large deflections of a laminate including the transverse shear effect, from which the stress equations of motion and traction boundary conditions can be written readily as before. Comments made earlier in connection with the classical Eq. (8.3.3) remain valid.

The plate stress–strain relations for the membrane forces and bending moments are the same as in Eqs. (8.3.6). In addition, however, the transverse shearing forces now must be included. These have the form

$$\begin{bmatrix} Q_y \\ Q_x \end{bmatrix} = \kappa \begin{bmatrix} A_{44} & A_{45} \\ A_{45} & A_{55} \end{bmatrix} \begin{bmatrix} 2\epsilon_{zy}^{(0)} \\ 2\epsilon_{zx}^{(0)} \end{bmatrix}, \tag{8.3.12}$$

where

$$[Q_x,\ Q_y] = \sum_{k=1}^{n} \int_{h_{k-1}}^{h_k} [\sigma_{zx(k)},\ \sigma_{zy(k)}]\, dz,$$

$A_{44}, \ldots$ are elements of A_{ij} above, and κ is a shear factor. In our early work (1959a, 1959b), the value of the shear correction factor was shown to be close to one for an ordinary sandwich plate. Here, we propose to adopt the value of one for the shear correction factor for a laminated composite plate, as in the linear case in Chapter 5.

In the simplified nonlinear case of small strains and large rotations, we have the following three- dimensional strain–displacement relations:

$$\epsilon_{xx} = \frac{\partial u_x}{\partial x} + \frac{1}{2}\omega_{zx}^2$$

$$\epsilon_{yy} = \frac{\partial u_y}{\partial y} + \frac{1}{2}\omega_{yz}^2$$

$$\epsilon_{xy} = \frac{1}{2}\left(\frac{\partial u_x}{\partial y} + \frac{\partial u_y}{\partial x} - \omega_{zx}\omega_{yz}\right)$$

$$\epsilon_{zx} = \frac{1}{2}\left(\frac{\partial u_x}{\partial z} + \frac{\partial u_z}{\partial x}\right)$$

$$\epsilon_{zy} = \frac{1}{2}\left(\frac{\partial u_y}{\partial z} + \frac{\partial u_z}{\partial y}\right),$$

where the first three are the same as in Eqs. (8.3.7). By virtue of Eqs. (8.3.10) and (8.3.2), the nonlinear plate strains of the zeroth and first order are now

$$\epsilon_{xx}^{(0)} = \frac{\partial u}{\partial x} + \frac{1}{2}\left(\frac{\partial w}{\partial x}\right)^2$$

$$\epsilon_{yy}^{(0)} = \frac{\partial v}{\partial y} + \frac{1}{2}\left(\frac{\partial w}{\partial y}\right)^2$$

$$\epsilon_{xy}^{(0)} = \frac{1}{2}\left(\frac{\partial u}{\partial y} + \frac{\partial v}{\partial x} + \frac{\partial w}{\partial x}\frac{\partial w}{\partial y}\right)$$

$$\epsilon_{xx}^{(1)} = \frac{\partial \psi}{\partial x}$$

$$\epsilon_{yy}^{(1)} = \frac{\partial \phi}{\partial y} \qquad\qquad (8.3.13)$$

$$\epsilon_{xy}^{(1)} = \frac{1}{2}\left(\frac{\partial \psi}{\partial y} + \frac{\partial \phi}{\partial x}\right)$$

$$\epsilon_{zx}^{(0)} = \frac{1}{2}\left(\psi + \frac{\partial w}{\partial x}\right)$$

$$\epsilon_{zy}^{(0)} = \frac{1}{2}\left(\phi + \frac{\partial w}{\partial y}\right).$$

The plate stress–displacement relations then are obtained by substituting Eqs. (8.3.13) into the plate stress–strain relations in Eqs. (8.3.6) and (8.3.12). By means of these results, the stress equations of motion can be transformed readily into the displacement equations of motion in the usual manner.

8.4 Nonlinear Vibration of an Orthotropic Symmetric Laminate

In the preceding section, we derived the stress equations of motion and stress–strain– displacement relations in both classical and refined theories of a laminate. In this section, we consider the simpler case of a specially orthotropic symmetric laminate for which $A_{16} = A_{26} = D_{16} = D_{26} = B_{ij} = \rho^{(1)} = 0$. By examining nonlinear free vibration, we recognize again the greater variety of frequencies generated by anisotropy (Yu 1993). One of our main concerns is still to assess the importance of transverse shear effect, thereby to help decide whether the classical or refined equations should be used in an analysis.

8.4.1 Displacement Equations of Motion

From the variational equations of motion (8.3.3), together with the stress–strain–displacement relations, the following three displacement equations of motion of the classical type are obtained:

$$
A_{11} \frac{\partial}{\partial x} \left[\frac{\partial u}{\partial x} + \frac{1}{2} \left(\frac{\partial w}{\partial x} \right)^2 \right] + A_{12} \frac{\partial}{\partial x} \left[\frac{\partial v}{\partial y} + \frac{1}{2} \left(\frac{\partial w}{\partial y} \right)^2 \right]
$$

$$
+ A_{66} \frac{\partial}{\partial y} \left(\frac{\partial u}{\partial y} + \frac{\partial v}{\partial x} + \frac{\partial w}{\partial x} \frac{\partial w}{\partial y} \right) + P_x^{(0)} + f_x^{(0)} - \rho^{(0)} \ddot{u} = 0
$$

$$
A_{12} \frac{\partial}{\partial y} \left[\frac{\partial u}{\partial x} + \frac{1}{2} \left(\frac{\partial w}{\partial x} \right)^2 \right] + A_{22} \frac{\partial}{\partial y} \left[\frac{\partial v}{\partial y} + \frac{1}{2} \left(\frac{\partial w}{\partial y} \right)^2 \right]
$$

$$
+ A_{66} \frac{\partial}{\partial x} \left(\frac{\partial u}{\partial y} + \frac{\partial v}{\partial x} + \frac{\partial w}{\partial x} \frac{\partial w}{\partial y} \right) + P_y^{(0)} + f_y^{(0)} - \rho^{(0)} \ddot{v} = 0
$$

$$
D_{11} \frac{\partial^4 w}{\partial x^4} + 2(D_{12} + 2D_{66}) \frac{\partial^4 w}{\partial x^2 \partial y^2} + D_{22} \frac{\partial^4 w}{\partial y^4}
$$

$$
- \frac{\partial}{\partial x} \left\{ A_{11} \left[\frac{\partial u}{\partial x} + \frac{1}{2} \left(\frac{\partial w}{\partial x} \right)^2 \right] \frac{\partial w}{\partial x} + A_{12} \left[\frac{\partial v}{\partial y} + \frac{1}{2} \left(\frac{\partial w}{\partial y} \right)^2 \right] \frac{\partial w}{\partial x} \right.
$$

$$
\left. + A_{66} \left[\frac{\partial u}{\partial y} + \frac{\partial v}{\partial x} + \frac{\partial w}{\partial x} \frac{\partial w}{\partial y} \right] \frac{\partial w}{\partial y} \right\} \tag{8.4.1}
$$

$$
- \frac{\partial}{\partial y} \left\{ A_{12} \left[\frac{\partial u}{\partial x} + \frac{1}{2} \left(\frac{\partial w}{\partial x} \right)^2 \right] \frac{\partial w}{\partial y} + A_{22} \left[\frac{\partial v}{\partial y} + \frac{1}{2} \left(\frac{\partial w}{\partial y} \right)^2 \right] \frac{\partial w}{\partial y} \right.
$$

$$
\left. + A_{66} \left[\frac{\partial u}{\partial y} + \frac{\partial v}{\partial x} + \frac{\partial w}{\partial x} \frac{\partial w}{\partial y} \right] \frac{\partial w}{\partial x} \right\}
$$

$$
- \frac{\partial P_x^{(1)}}{\partial x} - \frac{\partial P_y^{(1)}}{\partial y} - P_z^{(0)}
$$

$$- \frac{\partial f_x^{(1)}}{\partial x} - \frac{\partial f_y^{(1)}}{\partial y} - f_z^{(0)} + \rho^{(0)} \ddot{w} - \rho^{(2)} \nabla^2 \ddot{w} = 0.$$

Similarly, from the variational equation of motion (8.3.11), five displacement equations of motion of the refined type are obtained. The first two are the same as those in Eqs. (8.4.1). The remaining three are as follows:

$$
\begin{aligned}
& D_{11} \frac{\partial^2 \psi}{\partial x^2} + D_{12} \frac{\partial^2 \phi}{\partial x \partial y} + D_{66} \left(\frac{\partial^2 \psi}{\partial y^2} + \frac{\partial^2 \phi}{\partial x \partial y} \right) \\
& \quad - \kappa A_{55} \left(\psi + \frac{\partial w}{\partial x} \right) - \kappa A_{45} \left(\phi + \frac{\partial w}{\partial y} \right) + P_x^{(1)} + f_x^{(1)} - \rho^{(2)} \ddot{\psi} = 0 \\
& D_{12} \frac{\partial^2 \psi}{\partial x \partial y} + D_{22} \frac{\partial^2 \phi}{\partial y^2} + D_{66} \left(\frac{\partial^2 \psi}{\partial x \partial y} + \frac{\partial^2 \phi}{\partial x^2} \right) \\
& \quad - \kappa A_{45} \left(\psi + \frac{\partial w}{\partial x} \right) - \kappa A_{44} \left(\phi + \frac{\partial w}{\partial y} \right) + P_y^{(1)} + f_y^{(1)} - \rho^{(2)} \ddot{\phi} = 0
\end{aligned}
$$

$$
\begin{aligned}
& \frac{\partial}{\partial x} \left\{ \kappa A_{55} \left(\psi + \frac{\partial w}{\partial x} \right) + \kappa A_{45} \left(\phi + \frac{\partial w}{\partial y} \right) \right. \\
& \quad + A_{11} \left[\frac{\partial u}{\partial x} + \frac{1}{2} \left(\frac{\partial w}{\partial x} \right)^2 \right] \frac{\partial w}{\partial x} + A_{12} \left[\frac{\partial v}{\partial y} + \frac{1}{2} \left(\frac{\partial w}{\partial y} \right)^2 \right] \frac{\partial w}{\partial x} \\
& \quad \left. + A_{66} \left[\frac{\partial u}{\partial y} + \frac{\partial v}{\partial x} + \frac{\partial w}{\partial x} \frac{\partial w}{\partial y} \right] \frac{\partial w}{\partial y} \right\} \\
& + \frac{\partial}{\partial y} \left\{ \kappa A_{45} \left(\psi + \frac{\partial w}{\partial x} \right) + \kappa A_{44} \left(\phi + \frac{\partial w}{\partial y} \right) \right. \\
& \quad + A_{12} \left[\frac{\partial u}{\partial x} + \frac{1}{2} \left(\frac{\partial w}{\partial x} \right)^2 \right] \frac{\partial w}{\partial y} + A_{22} \left[\frac{\partial v}{\partial y} + \frac{1}{2} \left(\frac{\partial w}{\partial y} \right)^2 \right] \frac{\partial w}{\partial y} \\
& \quad \left. + A_{66} \left[\frac{\partial u}{\partial y} + \frac{\partial v}{\partial x} + \frac{\partial w}{\partial x} \frac{\partial w}{\partial y} \right] \frac{\partial w}{\partial x} \right\} \\
& + P_z^{(0)} + f_z^{(0)} - \rho^{(0)} \ddot{w} = 0.
\end{aligned}
\tag{8.4.2}
$$

When the nonlinear terms are neglected, Eqs. (8.4.1) and (8.4.2) reduce to the linear equations given earlier in Chapter 5.

8.4.2 Nonlinear Free Vibration

In the analysis of nonlinear free vibration, surface traction and body force terms in the displacement equations are ignored. With our emphasis on low frequencies, some inertia terms may also be neglected, and only the term $\rho^{(0)} \ddot{w}$ need be retained. By choosing to examine plane strain vibrations involving x-dependence only on the basis of the refined theory, the first two of Eqs. (8.4.1), together with the three

Eqs. (8.4.2), reduce to the following three:

$$A_{11}\left(\frac{\partial^2 u}{\partial x^2} + \frac{\partial w}{\partial x}\frac{\partial^2 w}{\partial x^2}\right) = 0 \tag{8.4.3}$$

$$D_{11}\frac{\partial^2 \psi}{\partial x^2} - \kappa A_{55}\left(\psi + \frac{\partial w}{\partial x}\right) = 0 \tag{8.4.4}$$

$$\kappa A_{55}\left(\frac{\partial \psi}{\partial x} + \frac{\partial^2 w}{\partial x^2}\right) + A_{11}\left[\frac{\partial u}{\partial x} + \frac{1}{2}\left(\frac{\partial w}{\partial x}\right)^2\right]\frac{\partial^2 w}{\partial x^2} - \rho^{(0)}\ddot{w} = 0. \tag{8.4.5}$$

For a laminate with immovable hinged edges at $x = 0,\ \ell$, we assume

$$w(x, t) = W\,\tau(t)\sin\frac{\pi x}{\ell}, \qquad \psi(x, t) = \Psi\,\tau(t)\cos\frac{\pi x}{\ell}. \tag{8.4.6}$$

Substitution of w into Eq. (8.4.3) and subsequent integration yield

$$u = -W^2\tau^2\,\frac{\pi}{8\ell}\,\sin\frac{2\pi x}{\ell}, \tag{8.4.7}$$

where the integration constants are set equal to zero to satisfy the conditions of immovable edges. Next, by substituting Eqs. (8.4.6) into (8.4.4) and by introducing a dimensionless parameter β, we find

$$\Psi = -\frac{1}{[1 + (D_{11}/\kappa A_{55})(\pi^2/\ell^2)]}\,\frac{\pi W}{\ell} = -\beta\,\frac{\pi W}{\ell}. \tag{8.4.8}$$

Finally, by virtue of Eqs. (8.4.6) through (8.4.8), Eq. (8.4.5) becomes

$$\rho^{(0)}W\ddot{\tau} + D_{11}\beta\,\frac{\pi^4}{\ell^4}\,W\tau + A_{11}\,\frac{\pi^4}{4\ell^4}\,W^3\tau^3 = 0. \tag{8.4.9}$$

We next suppress the transverse shear effect. For vibrations involving only the x-dependence, the classical equations (8.4.1) are reduced similarly as before, and Eq. (8.4.3) is still obtained. Instead of Eqs. (8.4.4) and (8.4.5), we now have the single equation

$$D_{11}\frac{\partial^4 w}{\partial x^4} - A_{11}\left[\frac{\partial u}{\partial x} + \frac{1}{2}\left(\frac{\partial w}{\partial x}\right)^2\right]\frac{\partial^2 w}{\partial x^2} + \rho^{(0)}\ddot{w} = 0. \tag{8.4.10}$$

By assuming the same form of w as in Eqs. (8.4.6), the result of u in Eq. (8.4.7) is again obtained. The final equation derived from Eq. (8.4.10) differs only slightly from Eq. (8.4.9) in that the parameter β now takes the value of unity. This is to be expected since suppression of the transverse shear effect can be effected by letting the shear correction factor κ equal to infinity, which makes β equal to 1.

We recognize that the governing Eq. (8.4.9) is a Duffing equation, from which the nonlinear frequency can be determined readily. Denoting the linear and nonlinear frequencies of the system by ω_0 and ω, respectively, and making use of the Duffing method, we find the ratio

$$\frac{\omega}{\omega_0} = \sqrt{1 + \frac{3}{16\beta} \frac{A_{11}}{D_{11}} W^2} \qquad \text{where} \quad \beta = \frac{1}{1 + \frac{D_{11}}{\kappa A_{55}} \frac{\pi^2}{\ell^2}}.$$

Vibrations involving only y-dependence may be treated similarly, the result being

$$\frac{\omega}{\omega_0} = \sqrt{1 + \frac{3}{16\beta} \frac{A_{22}}{D_{22}} W^2} \qquad \text{where} \quad \beta = \frac{1}{1 + \frac{D_{22}}{\kappa A_{44}} \frac{\pi^2}{\ell^2}}.$$

The situation is therefore the same as in the linear case in that anisotropy has generated more varieties and options for the frequencies. The above expressions of β show that the transverse shear effect becomes negligible if

$$\frac{D_{11}}{\kappa A_{55}} \frac{\pi^2}{\ell^2} \ll 1, \qquad \frac{D_{22}}{\kappa A_{44}} \frac{\pi^2}{\ell^2} \ll 1,$$

which also have been observed in the linear case. The nonlinear results have the typical form that shows an increase of the nonlinear frequency with the amplitude W. Finally, we observe that, since A_{11} and A_{22} are proportional to the thickness of the laminate and D_{11} and D_{22} are proportional to the cube of the thickness, the effect of large deflection becomes more important as the thickness decreases.

8.5   Equations for Large Deflections of a Sandwich Beam with Laminated Composite Facings and an Orthotropic Core

In this section, we combine the use of a sandwich and laminated composite by considering large deflections of a sandwich with laminated composite face layers and an orthotropic core. For simplicity, a beam is considered. On the other hand, to make this section self-contained, some earlier materials are repeated, but only slightly. The development follows our recent work (Yu 1992b) and is an extension of the treatment of a nonlinear Timoshenko beam in Section 7.1.

As before, the beam is taken to have an axis in the x-direction with the ends at $x = 0, \ell$. The beam width is equal to 1 so that this discussion also is applicable to a plate in plane strain. The middle plane of the beam or plate is in the xy-plane. Also as before, in the z-direction the thicknesses of the lower facing, core, and upper facing extend from $-h$ to $-h_1$, $-h_1$ to h_1, and h_1 to h, respectively. The thickness of the core is then $2h_1$, that of each facing is $h_2 = h - h_1$, and that of the entire sandwich is $2h$. While the two facings are made of the same materials, the core is generally made of a different material. In an ordinary sandwich, the facings have higher density and stiffness than the core. The lower facing, core, and upper facing are designated by the subscripts 1, 2, and 3, respectively. When there is no need to distinguish between the two facings, both are referred to by a single subscript 2.

8.5.1 Stress Equations of Motion

For the sandwich beam, we assume as before a perfect bond between adjacent layers by taking the displacements in the form

$$u_1 = u + z\psi, \qquad u_2, u_3 = u \mp h_1\psi$$
$$v_i = 0, \qquad w_i = w \qquad (i = 1, 2, 3) \tag{8.5.1}$$

in which continuity at the interfaces between adjacent layers is assured and the beam displacements u, w, and ψ are functions of x and t. We further take the rotations in the form

$$\omega_{yzi} = 0, \qquad \omega_{zxi} = -\frac{\partial w}{\partial x} \qquad (i = 1, 2, 3). \tag{8.5.2}$$

Equations (8.5.1) and (8.5.2) are substituted into the pseudo-variational equation (1.4.8), and integration is carried out with respect to z over the thickness of each of the layers. After combining the results for all layers by enforcing the continuity of tractions at the interfaces, the final result for the sandwich beam is found to be

$$\int_{t_0}^{t_1} \int_0^\ell \left(\left\{ \frac{\partial}{\partial x}(N_{x1} + N_{x2} + N_{x3}) - 2(\rho_1 h_1 + \rho_2 h_2)\ddot{u} \right\} \delta u \right.$$

$$+ \left\{ \frac{\partial}{\partial x}\left[Q_{x1} + (N_{x1} + N_{x2} + N_{x3})\frac{\partial w}{\partial x} \right] + p_z - 2(\rho_1 h_1 + \rho_2 h_2)\ddot{w} \right\} \delta w$$

$$\left. + \left\{ \frac{\partial}{\partial x}[M_{x1} + h_1(N_{x3} - N_{x2})] - Q_{x1} - \frac{2}{3}\rho_1 h_1^3 \ddot{\psi} \right\} \delta\psi \right) dx\, dt$$

$$- \int_{t_0}^{t_1} \left\{ [N_{x1} + N_{x2} + N_{x3} - p_x^{(0)}] \delta u \right.$$

$$+ \left[Q_{x1} + (N_{x1} + N_{x2} + N_{x3})\frac{\partial w}{\partial x} - p_z^{(0)} \right] \delta w$$

$$\left. + [M_{x1} + h_1(N_{x3} - N_{x2}) - p_x^{(1)}] \delta\psi \right\}_{x=0}^{x=\ell} dt = 0. \tag{8.5.3}$$

In this result, the body forces have been neglected and the beam stresses and tractions are

$$(N_{xi}, Q_{x1}, M_{x1}) = \int (\sigma_{xxi}, \sigma_{zx1}, \sigma_{xx1}z)\, dz$$
$$(p_x^{(0)}, p_z^{(0)}, p_x^{(1)}) = \int (p_x, p_z, p_x z)\, dz \tag{8.5.4}$$

in which integrations cover the appropriate thicknesses of the layers.

Equation (8.5.3) is the pseudo-variational equation of motion of the sandwich beam or plate, from which stress equations of motion and boundary conditions can be written in the usual manner. The core of an ordinary sandwich is usually so much weaker than the facings that we take

$$N_{x1} = M_{x1} = 0.$$

The stress equations of motion are then

$$(N_{x2} + N_{x3})' - 2(\rho_1 h_1 + \rho_2 h_2)\ddot{u} = 0$$
$$[Q_{x1} + (N_{x2} + N_{x3})w']' + p_z - 2(\rho_1 h_1 + \rho_2 h_2)\ddot{w} = 0 \qquad (0 \leq x \leq \ell)$$
$$h_1(N_{x3} - N_{x2})' - Q_{x1} - \tfrac{2}{3}\rho_1 h_1^3 \ddot{\psi} = 0$$

$$(8.5.5)$$

and the boundary conditions are

$$N_{x2} + N_{x3} = p_x^{(0)} \quad \text{or} \quad u = \text{prescribed}$$
$$Q_{x1} + (N_{x2} + N_{x3})w' = p_z^{(0)} \quad \text{or} \quad w = \text{prescribed} \qquad (8.5.6)$$
$$h_1(N_{x3} - N_{x2}) = p_x^{(1)} \quad \text{or} \quad \psi = \text{prescribed}.$$

In Eqs. (8.5.5) and (8.5.6), the sum $(N_{x2} + N_{x3})$ is the resultant axial force, its component $(N_{x2} + N_{x3})w'$ combined with Q_{x1} gives the total shearing force, and the product $h_1(N_{x2} - N_{x3})$ serves as the bending moment. Since these equations do not depend on material properties, they are valid for a sandwich with either isotropic or laminated composite face layers.

8.5.2   Displacements, Strains, and Stresses

Axial Forces in Isotropic Facings

For an isotropic elastic layer in a state of plane stress, the *nonlinear* stress–strain relations have the form

$$\sigma_{xxi} = \frac{E_2}{1 - v_2^2}(\epsilon_{xxi} + v_2 \epsilon_{yyi})$$
$$\sigma_{yyi} = \frac{E_2}{1 - v_2^2}(\epsilon_{yyi} + v_2 \epsilon_{xxi}) \qquad (i = 2, 3) \qquad (8.5.7)$$
$$\sigma_{xyi} = 2G_2 \epsilon_{xyi}.$$

In the plane strain case in which only the nonlinear strain ϵ_{xxi} exists, the first of these relations reduces to

$$\sigma_{xxi} = \frac{E_2 \epsilon_{xxi}}{1 - v_2^2} \qquad (i = 2, 3). \qquad (8.5.8)$$

This is the only stress needed for deriving N_{x2} and N_{x3}. Corresponding to the strain ϵ_{xxi}, the stress σ_{yyi} also will be present, but σ_{xyi} will not. For applications to a beam, Eq. (8.5.8) is simply replaced by

$$\sigma_{xxi} = E_2 \epsilon_{xxi} \qquad (i = 2, 3). \tag{8.5.9}$$

In the simplified nonlinear case of small strains and large rotations, the nonlinear strain component ϵ_{xxi} has the form

$$\epsilon_{xxi} = u_i' + \frac{1}{2}\omega_{zxi}^2. \tag{8.5.10}$$

For membrane facings, we substitute u_2, u_3 from Eqs. (8.5.1) and ω_{zx2}, ω_{zx3} from Eqs. (8.5.2) into Eqs. (8.5.10), (8.5.8), and (8.5.4), in that order. The final results are, for the facings in plane strain,

$$
\begin{aligned}
(N_{x2}, N_{x3}) &= \frac{E_2 h_2}{1 - v_2^2} \left(\epsilon_{xx2}^{(0)}, \epsilon_{xx3}^{(0)} \right) \\
&= \frac{E_2 h_2}{1 - v_2^2} \left[u' \mp h_1 \psi' + \tfrac{1}{2}(w')^2 \right],
\end{aligned}
\tag{8.5.11}
$$

where $\epsilon_{xx2}^{(0)}$, $\epsilon_{xx3}^{(0)}$ are nonlinear normal strains of zeroth order, uniform over the facing thickness. Similar results for a beam are obtained by simply replacing $(1 - v_2^2)$ by 1, but N_{x2}, N_{x3} are still forces per unit length in the direction of the beam width. To obtain the total axial force in a beam, these must be multiplied further by the width of the beam.

Axial Forces in Laminated Composite Facings

In the general nonlinear case of a classical multidirectional laminate, the relations between the plate stresses and plate strains are given by Eq. (8.3.6). When the laminate is symmetric, the equation reduces to

$$
\begin{bmatrix} N_x \\ N_y \\ N_{xy} \end{bmatrix} = \begin{bmatrix} A_{11} & A_{12} & A_{16} \\ A_{21} & A_{22} & A_{26} \\ A_{61} & A_{62} & A_{66} \end{bmatrix} \begin{bmatrix} \epsilon_{xx}^{(0)} \\ \epsilon_{yy}^{(0)} \\ 2\epsilon_{xy}^{(0)} \end{bmatrix}. \tag{8.5.12}
$$

To apply Eqs (8.5.12) to the facings of the sandwich, we insert as before a subscript $i = 2$ or 3 to all quantities. Since $\epsilon_{xxi}^{(0)}$ is the only nonzero strain in these equations, we find

$$(N_{x2}, N_{x3}) = (A_{11})_2 \left(\epsilon_{xx2}^{(0)}, \epsilon_{xx3}^{(0)} \right). \tag{8.5.13}$$

Comparing Eqs. (8.5.13) with (8.5.11), we see that $(A_{11})_2$ has taken the place of $E_2 h_2/(1 - v_2^2)$. These equations therefore may be written in a common form as follows:

$$(N_{x2}, N_{x3}) = K_2(\epsilon_{xx2}^{(0)}, \epsilon_{xx3}^{(0)}) = K_2 \left[u' \mp h_1 \psi' + \tfrac{1}{2}(w')^2 \right] \tag{8.5.14}$$

with

$$K_2 = \frac{E_2 h_2}{1 - v_2^2} \qquad \text{for isotropic facings}$$

$$= (A_{11})_2 \qquad \text{for laminated composite facings.}$$

According to Eqs. (8.5.12), a nonzero $\epsilon_{xxi}^{(0)}$ by itself will induce an N_{yi} as well as N_{xi} due to Poisson's effect, just as in the case of isotropic facings. In addition, however, an N_{xyi} also will be induced unless A_{61} vanishes; such coupling between in-plane normal and shearing forces has no counterpart in the isotropic case.

Shearing Force in Isotropic or Orthotropic Core

This is simply

$$Q_{x1} = \kappa K_1(\psi + w'), \tag{8.5.15}$$

where

$$K_1 = 2G_1 h_1 \qquad \text{for an isotropic core}$$

$$= 2c_{55} h_1 \qquad \text{for an orthotropic core,}$$

κ is a shear factor and can be taken equal of 1 as before, G_1 is the shear modulus of an isotropic material, and c_{55} is the stiffness of an orthotropic material with respect to the zx-directions, denoted by the subscripts 55, as used before in A_{55}.

8.5.3   Displacement Equations of Motion

As in the earlier discussion of a nonlinear Timoshenko beam, we concentrate on the lateral motion of a sandwich beam by keeping only the inertia term in the lateral direction in the second of Eqs. (8.5.5). According to the reduced form of the first of Eqs. (8.5.5), the resultant axial force $(N_{x2} + N_{x3})$ is thus independent of x and uniform throughout the entire length of the beam. The sum of the two forces is found from Eqs. (8.5.14) and integrated over the length of the beam to yield

$$N_{x2} + N_{x3} = K_2 \frac{2}{\ell}\left[u_0 + \frac{1}{2}\int_0^\ell (w')^2\, dx\right], \tag{8.5.16}$$

where u_0 is the relative axial displacement between the two ends of the beam. By means of Eqs. (8.5.14) through (8.5.16), the second and third of Eqs. (8.5.5) become finally

$$2K_2 h_1^2 \psi''' + K_2 \frac{2}{\ell}\left[u_0 + \frac{1}{2}\int_0^\ell (w')^2\, dx\right] w'' + p_z = 2(\rho_1 h_1 + \rho_2 h_2)\ddot{w}$$

$$\tag{8.5.17}$$

$$\kappa K_1(\psi + w') = 2K_2 h_1^2 \psi''. \tag{8.5.18}$$

These are the displacement equations of motion for a sandwich beam or plate. Since the core can be isotropic or orthotropic, and the facings isotropic or composed of laminated composites, four different combinations are possible.

8.5.4   A Dynamical Model

Now that Eqs. (8.5.17) and (8.5.18) for a sandwich beam have a form similar to Eqs. (7.1.11) and (7.1.12) for a nonlinear Timoshenko beam, a simple dynamical model may be constructed for a simply supported sandwich beam in the same manner as before by adopting a single-mode approximation. The final result is

$$2(\rho_1 h_1 + \rho_2 h_2)W\ddot{\tau} + 2K_2 h_1^2 \frac{\pi^4}{\ell^4}\beta(1-r)W\tau$$

$$+ \frac{\pi^4}{2\ell^4}K_2 W^3 \tau^3 = P_z \sin(\omega t), \qquad (8.5.19)$$

where

$$\beta = \frac{1}{1 + \frac{2K_2}{\kappa K_1}\frac{\pi^2 h_1^2}{\ell^2}}.$$

Equation (8.5.19) is, of course, a Duffing equation.

References

Hyer, M.W., W.J. Anderson, and R.A. Scott. (1976) Nonlinear Vibrations of Three-Layer Beams of Viscoelastic Cores—I. Theory. *Journal of Sound and Vibrations*, Vol. 46, pp. 121– 136.

Hyer, M.W., W.J. Anderson, and R.A. Scott. (1978) Nonlinear Vibrations of Three-Layer Beams of Viscoelastic Cores—II. Experiment. *Journal of Sound and Vibrations*, Vol. 61, pp. 25–30.

Kovac, E.J., Jr., W.J. Anderson, and R.A. Scott. (1971) Forced Nonlinear Vibration of a Damped Sandwich Beam. *Journal of Sound and Vibrations*, Vol. 17, pp. 25–39.

Wu, C.I. and J.R. Vinson. (1969) On the Nonlinear Oscillations of Plates Composed of Composite Materials. *Journal of Composite Materials*, Vol. 3, pp. 548–561.

Yu, Y.Y. (1959a) A New Theory of Elastic Sandwich Plates—One-Dimensional Case. *Journal of Applied Mechanics*, Vol. 26, pp. 415–421.

Yu, Y.Y. (1959b) Simple Thickness-Shear Modes of Vibration of Infinite Sandwich Plates. *Journal of Applied Mechanics*, Vol. 26, pp. 679–681.

Yu, Y.Y. (1962) Nonlinear Flexural Vibrations of Sandwich Plates. *Journal of the Acoustical Society of America*, Vol. 34, pp. 1176–1183.

Yu, Y.Y. (1963) Application of Variational Equation of Motion to the Nonlinear Vibration Analysis of Homogeneous and Layered Plates and Shells. *Journal of Applied Mechanics*, Vol. 30, pp. 79–86.

Yu, Y.Y. (1989) Dynamics and Vibration of Layered Plates and Shells—A Perspective from Sandwiches to Laminated Composites. Bound volume for the 30th AIAA/ASME/ASCE/AHS/ASC Structures, Structural Dynamics and Materials Conference, pp. 2236–2245.

Yu, Y.Y. (1991) Nonlinear Dynamics and Chaos of Buckled Sandwich Beams with Isotropic or Laminated Composite Facings. In: *Proceedings of the Eighth International Conference for Composite Materials*, pp. 3-D-1 to 3-D-15.

Yu, Y.Y. (1992a) Dynamics and Vibration of Layered Plates and Shells—A Perspective from Sandwiches to Laminated Composites. Russian translation. In: *Mechanics of Composite Materials, Vol. 2, Construction of Composites, Proceedings of the First Soviet-American Symposium*, pp. 99–108. Riga, Latvia.

Yu, Y.Y. (1992b) Equations for Large Deflections of Homogeneous and Layered Beams with Application to Chaos and Acoustic Radiation. *Composites Engineering*, Vol. 2, pp. 117– 136.

Yu, Y.Y. (1992c) Equations for Large Deflections of Elastic and Piezoelectric Plates and Shallow Shells, Including Sandwiches and Laminated Composites, with Applications to Vibrations, Chaos, and Acoustical Radiation. Presented at the XIIIth International Congress of Theoretical and Applied Mechanics, Haifa, Israel.

Yu, Y.Y. (1992d) Generalized Variational Equations of Motion in Nonlinear Anisotropic Elasticity and Plate Theories. In: *Collection of Papers, a Volume Dedicated to the 80th Birthday of Academician V.V. Novozhilov*, edited by N.S. Solomenko, pp. 63–74. Shipbuilding Publishing House, St. Petersburg, Russia.

Yu, Y.Y. (1993) Nonlinear Large-Deflection Vibrations of Laminated Composite Plates. In: *Proceedings of the 8th Technical Conference on Composite Materials*, American Society for Composites, pp. 1099–1108. Technomic Publishing Co., Lancaster, Pennsylvania.

Yu, Y.Y. (1995a) On the Ordinary, Generalized, and Pseudo-Variational Equations of Motion in Nonlinear Elasticity, Piezoelectricity, and Classical Plate Theories. *Journal of Applied Mechanics*, Vol. 62, pp. 471–478.

Yu, Y.Y. (1995b) Some Recent Advances in Linear and Nonlinear Dynamical Modeling of Elastic and Piezoelectric Plates. *Journal of Intelligent Material Systems and Structures*, Vol. 6, pp. 237–254.

9

Chaotic Vibrations of Beams

In this chapter, we continue our discussion on nonlinear vibrations by concentrating on chaotic vibrations of beams. Study of chaos is an important branch of nonlinear vibrations that has attracted a great deal of attention in recent years as one of the fastest-growing disciplines in science, mathematics, and engineering. Early study of chaos in structural dynamics was associated with buckled elastic beams. Nonlinear vibrations of buckled beams were investigated by Eisley (1964) and by Tseng and Dugundji (1971). Chaos was observed experimentally and confirmed numerically by the latter authors, although the name was not mentioned. Later, chaos in beams buckled by nonlinear magnetic body forces was investigated by Moon and described in his books (1987, 1992). A systematic account of nonlinear dynamics and chaos has also been given in the books by Guckenheimer and Holmes (1983) and by Thompson and Stewart (1986).

Related materials that have been covered earlier include the nonlinear Timoshenko beam in Chapter 7 and the nonlinear sandwich beam in Chapter 8. Analysis of both cases has led to a dynamical model represented by a Duffing equation, which is one of the most studied in nonlinear dynamics and, more recently, in chaos. By means of this equation, chaos is covered in this chapter only briefly (Yu 1991, 1992). Among other things, we report some new numerical results on the effect of small damping on chaos for the transition between conservative and dissipative systems. These are presented in Sections 9.1 and 9.2. Chaos in beam vibrations is then treated in Section 9.3, and the related acoustic radiation is treated in Section 9.4.

9.1 A Numerical Study of Chaos According to Duffing's Equation: Effect of Damping

The phenomenon of chaos has generated much excitement since the 1960s and has been studied by physicists, mathematicians, and engineers, among others. The studies have covered both conservative Hamiltonian systems with zero damping and dissipative systems in which damping is intimately involved. These two types of systems are usually considered in the literature separately, although the contrast and similarity between their choatic behaviors can be quite interesting and revealing. This is demonstrated here through numerical analysis of Duffing's equation.

Duffing's equation is of particular importance to the dynamic analysis of many types of engineering structures undergoing nonlinear large deflections, as shown in earlier chapters. Although some damping always exists in real structures, the importance of zero damping is reflected in the fact that the theory of elasticity and its application to structural theory have always played an indispensable role in the training and practice of many engineers. This makes elastic systems a very important special case of Hamiltonian systems. In any case, a better knowledge of chaos exhibited by Duffing's equation is apparently still needed for both Hamiltonian and dissipative systems, and particularly for the transition between the two (Yu 1992). For instance, it will be important to be able to avoid chaos through the use of judiciously selected structural constructions and configurations, as we have suggested. A related task will be to properly construct the dynamical models of such configurations for studies of chaos.

For a variety of elastic structures, Duffing's equation takes the form

$$\ddot{x} + ax + x^3 = B \cos t, \tag{9.1.1}$$

which includes a forcing term on the right-hand side. A comprehensive numerical study was carried out by Ueda (1980) for chaos exhibited by the following form of the equation:

$$\ddot{x} + k\dot{x} + x^3 = B \cos t, \tag{9.1.2}$$

which includes a damping term but no longer the linear stiffness term. In fact, damping effect was included in all numerical results presented by Ueda. Thompson and Stewart (1986) further demonstrated chaos associated with Duffing's equation in the following more general form:

$$\ddot{x} + k\dot{x} + ax + x^3 = B \cos t, \tag{9.1.3}$$

where both damping and linear stiffness are included and the latter can be positive, zero, or negative. Results on chaos in the absence of damping according to the Duffing equation were first given by Yu (1991, 1992), with $a = 0$ in Eq. (9.1.1), with $k = 0$ in Eq. (9.1.2), or with $a = k = 0$ in Eq. (9.1.3).

In fact, based on the Duffing equation, we have looked into chaos for both zero and nonzero damping. Numerical results for both situations have been reported in our recent work (Yu 1992) and are reproduced here. Thus, Figure 9.1.1 shows

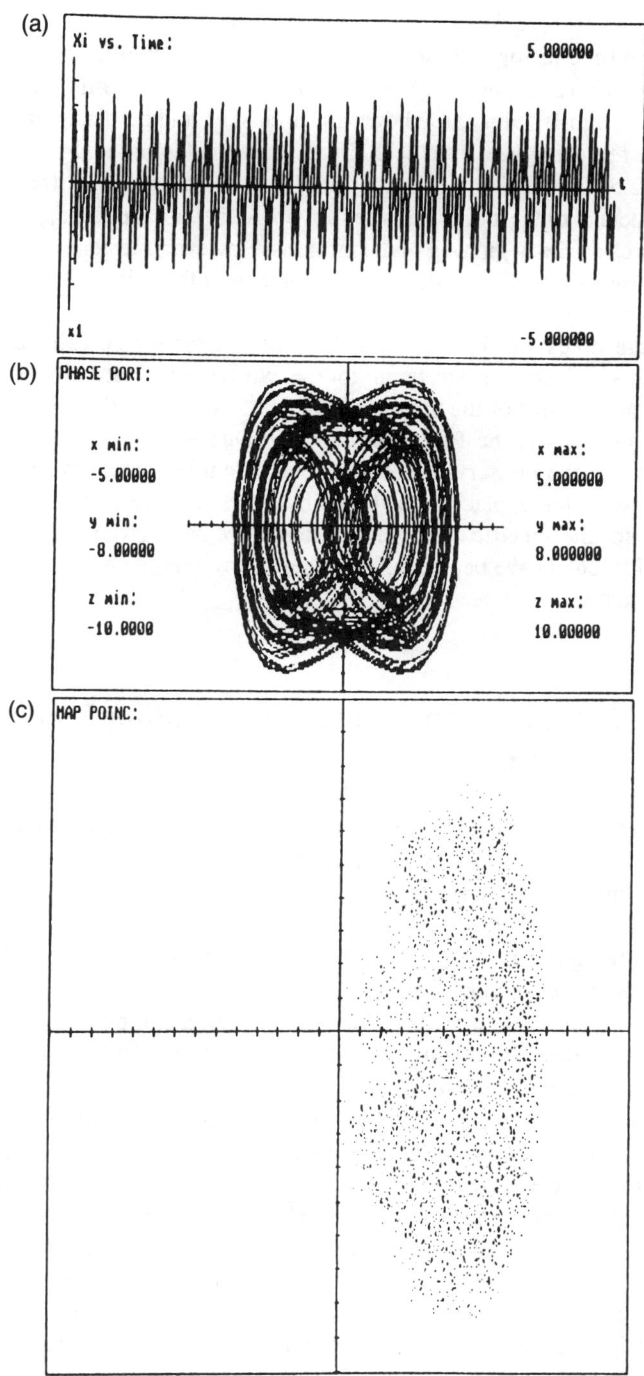

Fig. 9.1.1. A conservative system with $k = 0$, $a = 0$, $B = 7.5$, and zero initial conditions: (a) time history for $t = 0$–250; (b) phase portrait for $t = 0$–250; (c) Poincaré map for $t = 0$–20,000.

195

the plots of the time history, phase portrait, and Poincaré map for a conservative system with $k = 0$ (zero damping), $a = 0$, $B = 7.5$, and zero initial conditions. In Figure 9.1.2, similar plots are shown for a closely related dissipative system with $k = 0.05$ replacing $k = 0$. In each figure, the time history and phase portrait are for a small beginning part of the motion. Because damping is different in the two systems, details of their time histories are obviously not the same, particularly since they represent the beginning of the motions. However, the general outlook of the phase portraits of the two systems is similar, and both indicate the occurrence of chaos.

The Poincaré maps are for the entire time period for which calculation was carried out and are clearly very different for the two systems. Naturally, because of damping, the overall size of the map in Figure 9.1.2 is smaller than that in Figure 9.1.1. More importantly, the Poincaré map in Figure 9.1.1 exhibits completely random chaos for the conservative system, but Figure 9.1.2 shows the typical folding action that takes place in a dissipative system. The phase portrait and Poincaré map for the specific dissipative system have been given before by Ueda (1980), and his results have been confirmed here. However, he has not shown any result for zero damping.

9.2   More Poincaré Maps According to Duffing's Equation for Small Damping

The numerical study of chaos in the preceding section is continued in this section by increasing the damping from zero very gradually. The results are again based on the use of Duffing's equation for zero and small damping, and for positive, zero, and negative linear stiffness. Many more interesting Poincaré maps have been constructed. Among other things, these show that by adding very small damping to a Hamiltonian system originally exhibiting chaos, the chaos can disappear and the motion can become regular. Only when large enough damping is added can a strange attractor appear. Increasing the value of the linear stiffness has a similar effect as increasing the value of damping.

Twelve Poincaré maps constructed form Duffing's equation are shown in Figures 9.2.1 through 9.2.4. As before, the coefficient B in the equation is still taken equal to 7.5. Similarly, the coefficient k covers the same range between 0 and 0.05, but we let its value increase from 0 more gradually and systematically by taking k equal to 0.005, 0.025, 0.03, and 0.05. Finally, while the coefficient a was taken equal to only zero before, it is now also assigned the values $+0.2$ and -0.2. Each of the four figures is for a single value of k and includes three parts for the three values of a. All computations were carried out for the same initial conditions of $x = \dot{x} = 0$ at $t = 0$ and for the same time period from $t = 0$ to 16,000. The Poincaré maps were taken in the plane $z = 3$. The Poincaré map for $k = 0.05$ and $a = 0$ repeats the result in Figure 9.1.2 and also has been given before by Ueda (1980). Maps for $k = 0.05$ and $a = 2, -2$ have been given before by Thompson

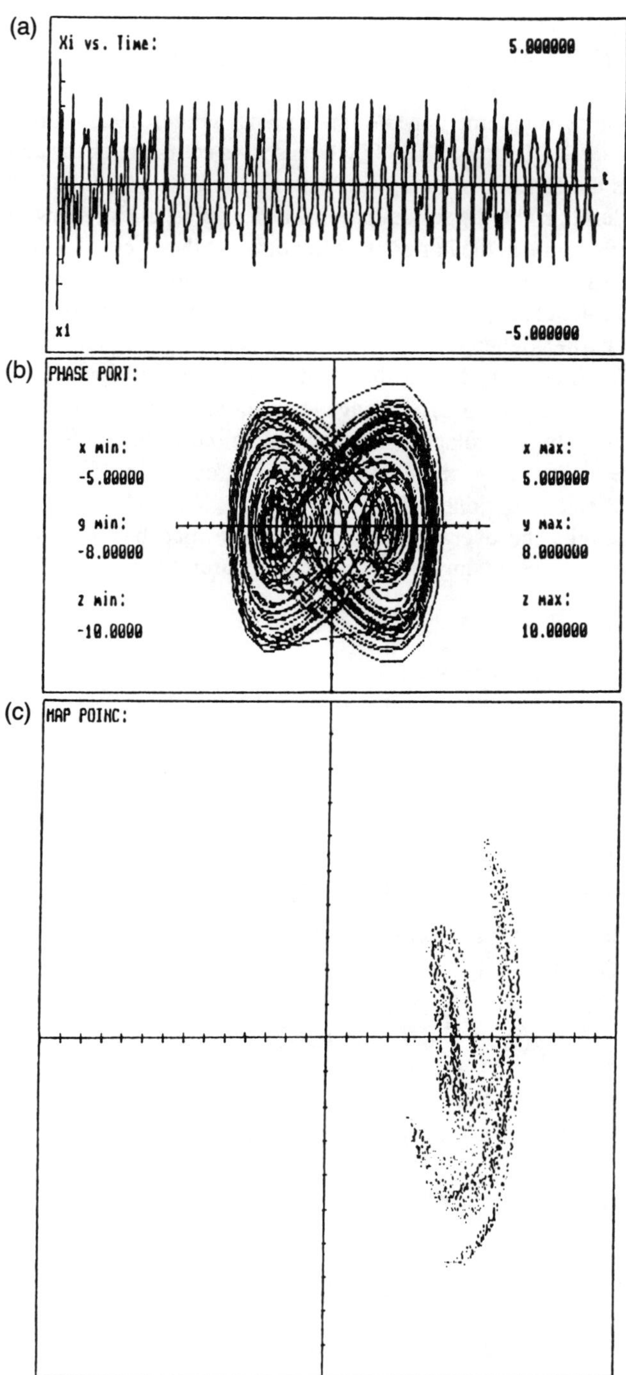

Fig. 9.1.2. A dissipative system with $k = 0.05$, $a = 0$, $B = 7.5$, and zero initial conditions: **(a)** time history for $t = 0$–250; **(b)** phase portrait for $t = 0$–250; **(c)** Poincaré map for $t = 0$–20,000.

197

and Stewart (1986). All other results in Figures 9.2.1 through 9.2.4 are new and have not been reported before.

It is important to supplement the Poincaré maps in Figures 9.2.1 through 9.2.4 by the one in Figure 9.1.1 for $k = a = 0$, which shows complete chaos and will go on forever due to the absence of damping. With Figures 9.2.1 through 9.2.4 laid side by side from left to right, the results provide a clear perspective showing the effect of increasing damping in the horizontal direction and the effect of increasing linear stiffness in the vertical direction. When small damping is added to a system that originally exhibits chaos, the chaos disappears and the motion becomes regular, as mentioned earlier. When the added damping is large enough, a strange attractor may appear. It is interesting to note that in some of these results, such as those given in Figures 9.2.2(c) and 9.2.3(c), a strange attractor appears to be in the making but is finally replaced by a regular motion. Increasing the value of the linear stiffness from negative to zero to positive has a similar effect as increasing the value of damping, as already mentioned.

Not surprisingly, the overall size of a Poincaré map becomes smaller when damping increases. Increasing linear stiffness apparently has a similar effect.

9.3   Spectral Analysis of Chaos

Frequency spectrum in spectral analysis is a valuable tool for the analysis of chaos (Yu 1992). It is based on the use of Fourier Transforms (FT), which will be reviewed first.

9.3.1   Fourier Transforms and Frequency Spectra

As an example, consider the velocity $v(t)$ and its fourier transform $V(\omega)$. The continuous FT pair are

$$v(t) = \frac{1}{2\pi} \int_{-\infty}^{\infty} V(\omega) \exp\left(-i\omega t\right) \, d\omega \qquad (9.3.1)$$

$$V(\omega) = \int_{-\infty}^{\infty} v(t) \exp\left(i\omega t\right) \, dt. \qquad (9.3.2)$$

If $v(t)$ is a periodic function with period T, it always can be written as a Fourier series in the form

$$v(t) = \sum_{n=-\infty}^{\infty} V_n \exp\frac{-i2\pi n t}{T}, \qquad (9.3.3)$$

where the complex coefficients are given by

$$V_n = \frac{1}{T} \int_0^T v(t) \exp\frac{i2\pi n t}{T} \, dt. \qquad (9.3.4)$$

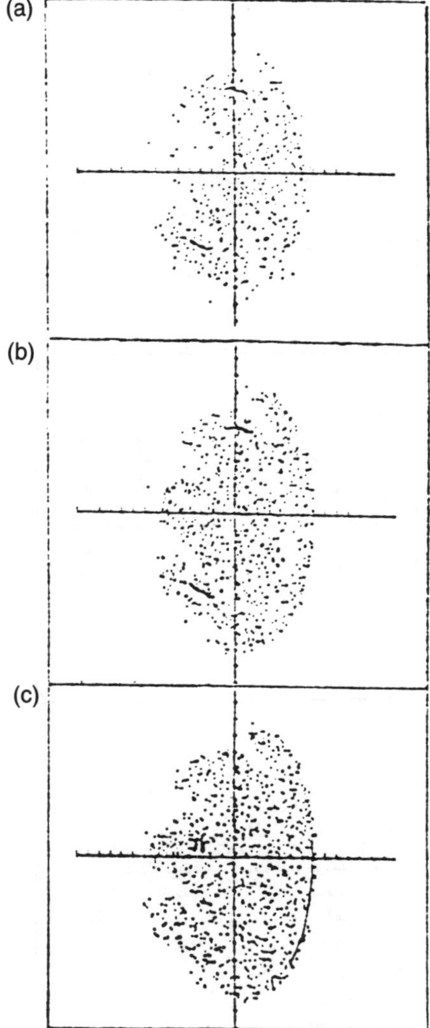

Fig. 9.2.1. Poincaré maps for $k = 0.005$: **(a)** $a = 0.2$; **(b)** $a = 0$; **(c)** $a = -0.2$.

Although the original periodic quantity $v(t)$ is a continuous function of time, it is often the case that only sampled values of the function are available, in the form of a discrete time series $\{v_r\}$. If N is the number of samples, all equally spaced with a spacing equal to $\Delta = T/N$, we can write

$$v_r = v_{t=r\Delta}, \qquad r = 0, 1, 2, \ldots, (N-1).$$

In fact, by taking N samples within a time span T at an equal time spacing Δ, an *aperiodic* function often is treated as if it were a periodic function with a

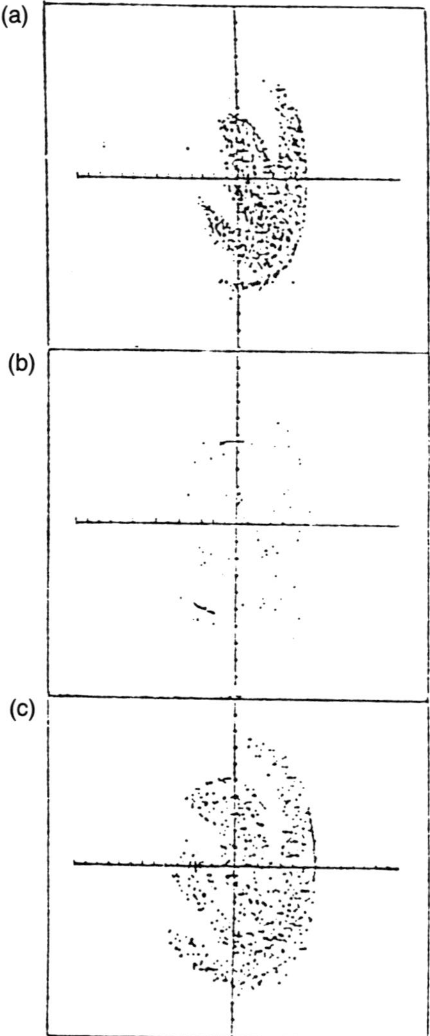

Fig. 9.2.2. Poincaré maps for $k = 0.025$: **(a)** $a = 0.2$; **(b)** $a = 0$; **(c)** $a = -0.2$.

period equal to the same T. If the integral in Eq. (9.3.4) now is approximated by the summation

$$V_n = \frac{1}{N} \sum_{r=0}^{N-1} v_r \exp \frac{i2\pi nr}{N} \qquad (n = 0, 1, 2, \ldots, N-1), \qquad (9.3.5)$$

then it can be shown that

$$v_r = \sum_{n=0}^{N-1} V_n \exp \frac{-i2\pi nr}{N} \qquad (r = 0, 1, 2, \ldots, N-1). \qquad (9.3.6)$$

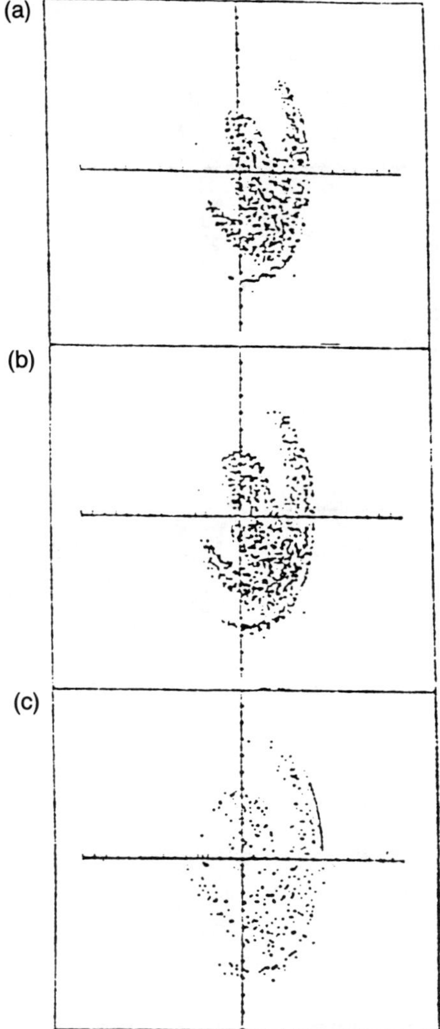

Fig. 9.2.3. Poincaré maps for $k = 0.03$: **(a)** $a = 0.2$; **(b)** $a = 0$; **(c)** $a = -0.2$.

Equations (9.3.5) and (9.3.6) are the Discrete Fourier Transform (DFT) pair for the discrete time series $\{v_r\}$. Even though Eq. (9.3.5) was formulated as an approximation to Eq. (9.3.4), the DFT pair are *exact* relations as far as the discrete values are concerned. As the time spacing Δ is made smaller and smaller, the DFT approach the FT as a limit.

The Fast Fourier Transform (FFT) is an ingenious computer algorithm for calculating the DFT on the basis of Eqs. (9.3.5) and (9.3.6). It is ingenious in that compared with straightforward calculations based on the same equations, computing time can be reduced by a huge factor equal to $N^2/N \log_2 N$. For instance, this

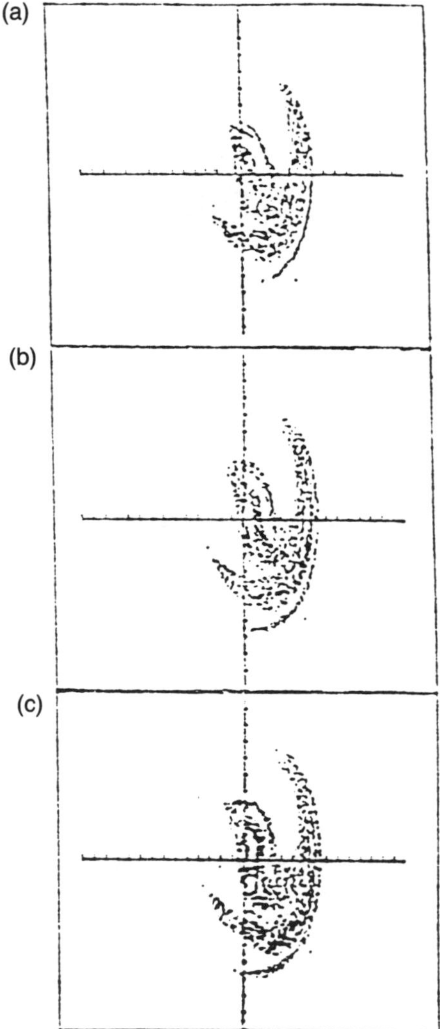

Fig. 9.2.4. Poincaré maps for $k = 0.05$: **(a)** $a = 0.2$; **(b)** $a = 0$; **(c)** $a = -0.2$.

factor is equal to 4096 for $N = 65,536$. Important contributions have been made on the FFT by Cooley and Tukey (1965), among others.

9.3.2   Application of Frequency Spectrum to Analysis of Chaos

The frequency spectrum can be used to identify the occurrence of chaos in vibrations. As is usually the case, it shows narrow spikes corresponding to the discrete harmonic components. If the motion becomes chaotic, a continuous frequency dis-

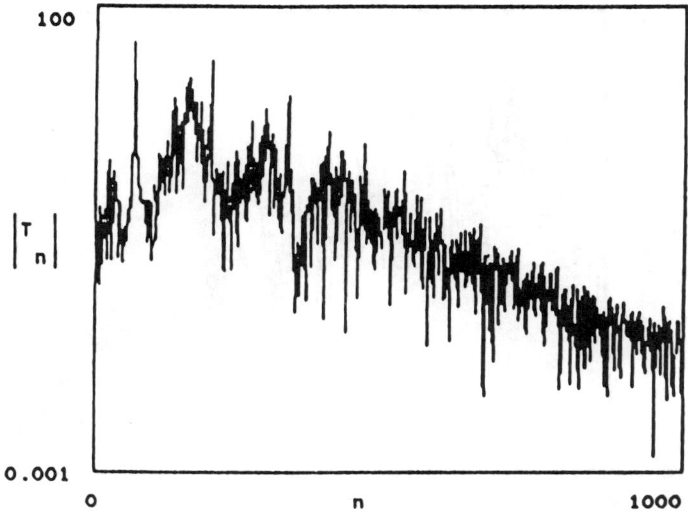

Fig. 9.3.1. Frequency spectrum of a conservative system with $k = 0$ (zero damping), $a = 0$, $B = 7.5$, and zero initial conditions.

tribution also appears. When the motion is fully chaotic, the discrete spikes may be overshadowed by the continuous spectrum.

As examples, the frequency spectra are shown in Figures 9.3.1 and 9.3.2 for the two numerical cases discussed earlier in Section 9.1 for zero and nonzero damping, respectively. The results were obtained through the use of the FFT for a time range of $t = 0$ to 500. The results show a combination of spikes and continuous frequency distributions for both cases, confirming the presence of chaos in addition to discrete harmonic components. They also show that damping lowers the overall level of the spectrum but has apparently caused little change in the general appearance. Furthermore, by expanding the time range from $t = 500$ to $t = 800$, the difference is essentially insignificant. This seems to indicate that the effect of the initial transient motion is small. The results also reflect the advantage of using the frequency spectrum to characterize chaos. It will be interesting to look further into different time slices and compare the results.

9.4   Acoustic Radiation from Chaotic Vibrations of a Beam

When a beam vibrates in a fluid medium and choas develops, acoustic radiation is expected to be generated in the usual manner. In this section, we propose a method of solution of the general problem, consisting of two steps that are assumed to be uncoupled from each other (Yu 1992). In the first step, the nonlinear and chaotic vibrations of a beam under harmonic excitation are first determined in the time domain and then analyzed in the frequency domain, for instance, by the FFT. In

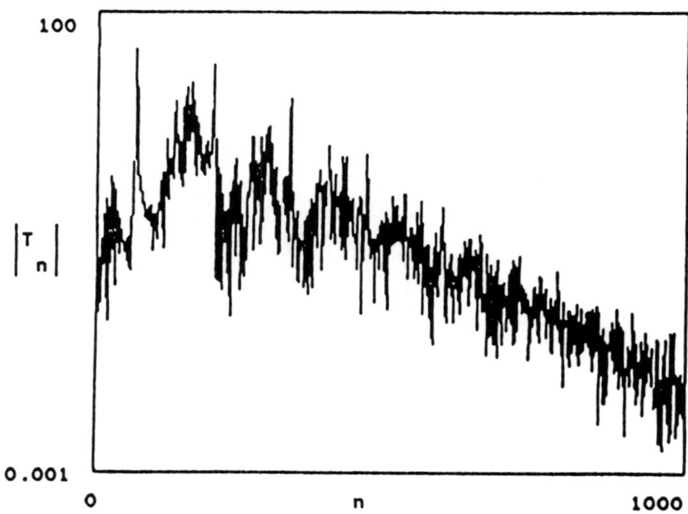

Fig. 9.3.2. Frequency spectrum of a dissipative system with $k = 0.05$ (nonzero damping), $a = 0$, $B = 7.5$, and zero initial conditions.

the second step, the acoustic radiation generated in the far field is analyzed as a linear problem on the basis of the classical wave equation by carrying out Fourier transforms in both time and space domains.

We consider a beam that has an axis along the x-axis and a unit width in the y-direction. It is in contact with a fluid medium that fills the space $z > 0$. During vibration, the deflection of the beam takes place in the z-direction and is a function of x and t, but not of y.

There are two equations that govern the acoustic pressure and fluid velocity, which will be denoted by $\mathbf{p}$ and $\mathbf{v}$, respectively. The first of these is the classical wave equation,

$$\frac{\partial^2 \mathbf{p}}{\partial x^2} + \frac{\partial^2 \mathbf{p}}{\partial z^2} = \frac{1}{c^2} \frac{\partial^2 \mathbf{p}}{\partial t^2}, \tag{9.4.1}$$

where c is the speed of sound in the fluid medium. The second equation is that of momentum as applied to the interface between the beam and fluid,

$$\left(\frac{\partial \mathbf{p}}{\partial z} \right)_{z=0} = -\rho \frac{\partial \mathbf{v}}{\partial t}, \tag{9.4.2}$$

where ρ is the mean density of the fluid and, being at $z = 0$, $\mathbf{v}$ has become the beam velocity. By writing the pressure and beam velocity in the general form

$$\mathbf{p} = \mathbf{p}(x, z, t), \qquad \mathbf{v} = \mathbf{v}(x, t), \tag{9.4.3}$$

the acoustic power radiated from an arbitrary length ℓ and a unit width of the beam is given by

$$P = \frac{1}{T} \int_0^T \int_0^\ell \text{Re} \{ \mathbf{p}(x, 0, t) \, \mathbf{v}^*(x, t) \} \, dx \, dt, \tag{9.4.4}$$

where T is the time span under consideration. We also have adopted the use of complex notations, with an asterisk denoting complex conjugate.

When the beam vibration is simple harmonic, the fluid velocity is in phase with the pressure in the far field at large distances from the source. Radiation due to simple harmonic vibrations of both an infinite beam and a finite beam has been well treated, and a detailed discussion has been given by Cremer and Heckl (1973, 1988). When the finite beam undergoes an arbitrary vibration, including any nonlinear and chaotic vibration, the time dependence need not be simple harmonic, and the general expressions of pressure and velocity in Eqs. (9.4.3) must be resorted to. We now introduce the following double FT pairs in both time and space domains:

$$p(x, z, t) = \left(\frac{1}{2\pi} \right)^2 \int_{-\infty}^{\infty} \int_{-\infty}^{\infty} P(k_x, \omega) \, \exp\left[-i(\omega t - k_x x) \right] dk_x \, d\omega$$

$$P(k_x, \omega) = \int_0^T \int_{-\infty}^{\infty} p(x, z, t) \, \exp\left[i(\omega t - k_x x) \right] dx \, dt$$

$$\tag{9.4.5}$$

$$v(x, t) = \left(\frac{1}{2\pi} \right)^2 \int_{-\infty}^{\infty} \int_{-\infty}^{\infty} V(k_x, \omega) \, \exp\left[-i(\omega t - k_x x) \right] dk_x \, d\omega$$

$$V(k_x, \omega) = \int_0^T \int_0^\ell v(x, t) \, \exp\left[i(\omega t - k_x x) \right] dx \, dt.$$

By means of these, Eqs. (9.4.1) and (9.4.2) are easily transformed. The transformed equations yield

$$P(k_x, \omega) = \frac{kc\rho}{(k^2 - k_x^2)^{1/2}} \, V(k_x, \omega) \, \exp\left[-i(k^2 - k_x^2)^{1/2} z \right], \tag{9.4.6}$$

where $k = \omega/c$ is the wave number. The radiated power is then, according to Eq. (9.4.4),

$$P = \frac{\rho}{4\pi^2 T} \int_{-\infty}^{\infty} \int_{-k}^{k} \frac{\omega |V(k_x, \omega)|^2}{(k^2 - k_x^2)^{1/2}} \, dk_x \, d\omega, \tag{9.4.7}$$

where the integration limits for k_x have reduced from $-\infty$ and ∞ to $-k$ and k, respectively, because the square root is not real for $k_x^2 > k^2$.

The dynamic model that has been proposed for a beam is based on the single-mode deflection form assumed in the first of Eqs. (7.1.13). The corresponding beam velocity is thus

$$v(x, t) = \dot{w}(x, t) = W \sin \frac{\pi x}{\ell} \dot{\tau}(t).$$

To separate the two parts that are functions of x and t, respectively, we rewrite this in the form

$$v(x, t) = w(x) \, \dot{\tau}(t)$$

so that the part that is a function of x is

$$w(x) = W \sin \frac{\pi x}{\ell}.$$

By next introducing the separate Fourier transforms

$$W(k_x) = \int_0^\ell w(x) \, \exp\,(ik_x x) \, dx$$

$$T(\omega) = \int_0^T \dot{\tau}(t) \, \exp\,(i\omega t) \, dt$$

for the two individual parts, it becomes clear that, from the last of Eqs. (9.4.5),

$$V(k_x, \omega) = T(\omega) \, W^*(k_x).$$

The acoustic power in Eq. (9.4.4) is therefore finally

$$P = \frac{\rho}{4\pi^2 T} \int_{-\infty}^\infty \int_{-k}^k \frac{\omega |T(\omega)|^2 |W(k_x)|^2}{(k^2 - k_x^2)^{1/2}} \, dk_x \, d\omega$$

References

Cooley, J.W. and J.W. Tukey. (1965) An Algorithm for the Machine Calculation of Complex Fourier Series. *Mathematical Computation*, Vol. 19, pp. 297–301.

Cremer, L. and M. Heckl. (1973) *Structure-Borne Sound*. Springer-Verlag, New York (2nd edition, 1988, with translation and revision by E.E. Unger).

Eisley, J.G. (1964) Large Amplitude Vibration of Buckled Beams and Rectangular Plates. *AIAA Journal*, Vol. 2, pp. 2207–2209.

Guckenheimer, J. and P. Holmes. (1983) *Nonlinear Oscillations, Dynamical Systems, and Bifurcations of Vector Fields*. Springer-Verlag, New York.

Moon, F.C. (1987) *Chaotic Vibrations*. Wiley, New York.

Moon, F.C. (1992) *Chaotic and Fractal Dynamics*. Wiley, New York.

Ott, E., C. Grebogi, and J.A. Yorke. (1990) Controlling Chaos. *Physical Review Letters*, Vol. 64, pp. 1196–1199.

Thompson, J.M.T. and J.B. Stewart. (1986) *Nonlinear Dynamics and Chaos*. Wiley, New York.

Tseng, W.Y. and J. Dugundji. (1971) Nonlinear Vibrations of a Buckled Beam under Harmonic Excitation. *Journal of Appllied Mechanics*, Vol. 38, pp. 467–476.

Ueda, Y. (1980) Steady Motions Exhibited by Duffing's Equation: A Picture Book of Regular and Chaotic Motions. In: *New Approaches to Nonlinear Problems in Dynamics*, edited by P.J. Holmes, pp. 311–322. SIAM, Philadelphia, Pennsylvania.

Yu, Y.Y. (1991) Nonlinear Dynamics and Chaos of Buckled Laminated Composite Beams with Isotropic or Laminated Composite Facings. In: *Proceedings of Eighth International Conference on Composite Materials*, pp. 3-D-1 to 3-D-15. SAMPE, Covina, California.

Yu, Y.Y. (1992) Equations for Large Deflections of Homogeneous and Layered Beams with Application to Chaos and Acoustic Radiation. *Composites Engineering*, Vol. 2, pp. 117–136.

10

Nonlinear Modeling of Piezoelectric Plates

In the three preceding chapters, we have discussed nonlinear dynamical modeling for large deflections of elastic beams, plates, and shallow shells, together with applications to nonlinear and chaotic vibrations. In this chapter, we further extend our discussion to nonlinear dynamical modeling for large deflections of piezoelectric plates.

At about the time when Mindlin published his first paper (1951a) on linear vibrations of isotropic elastic plates, he also presented his first results on high frequency linear vibrations of crystal plates (Mindlin 1951b) and piezoelectric plates (Mindlin 1952). As mentioned in Chapter 5, although not recognized in the literature, his equations of crystal plates turned out to have exactly the same form as the refined equations of flexure of symmetric laminates. Similarly, although the contributions made by Mindlin and his coworkers to high-frequency vibrations of linear piezoelectric crystal plates (Tiersten and Mindlin 1962; Mindlin 1972, 1974, 1984; Lee et al. 1987) always have been well received by physicists and electrical engineers, they are beginning to attract attention only recently from mechanical and structural engineers. This has been mainly through the book on linear piezoelectric plates by Tiersten (1969). The linear piezoelectric plate has also been treated by Bugdayci and Bogy (1981) and many others.

Vibration control of flexible structures long has been recognized as an important engineering discipline. At first, investigation was restricted to discrete active vibration control, such as was demonstrated by Yu (1968). More recently, new material systems have made it feasible to carry out active control of distributed parameter systems not only technically, but also economically. Studies in this and related areas have been made by Baily and Hubbard (1987), Lee and Moon (1990), Newnham and Ruschau (1991), Tzou (1991), Janas and Safari (1995), and others.

Among the new materials are the piezoelectric layers to be discussed in this chapter. Since a very rich literature on linear piezoelectric plates already exists, we choose to concentrate on the nonlinear dynamical modeling for large deflections of such plates. Thus, Section 10.1 serves as a link between nonlinear elasticity and piezoelectricity (Yu, 1992, 1995a,b,c). In Section 10.2, we extend the generalized Hamilton principle and variational equation of motion in three-dimensional nonlinear elasticity by further including the piezoelectric effect. Based on this newly extended version of the generalized variational equation of motion, the classical equations for large deflections of a piezoelectric plate are derived in Section 10.3, and the refined equations, including thickness effects, in Section 10.4. The latter represent the nonlinear counterpart of Mindlin's last set of linear equations for a piezoelectric plate. Some final remarks are offered in the last section.

10.1   From Elasticity to Piezoelectricity

In Chapter 1, we have characterized the classical nonlinear case in elasticity with *large finite deformations* and the simplified nonlinear cases with *small finite deformations*. While the deformations in all these cases are finite, the latter cases deal only with small strains, although large rotations are allowed to develop in thin structures. Thus, nonlinear elasticity is associated with finite deformations, in contrast to linear elasticity which is associated with *infinitesimal deformations*.

Small finite deformations make the simplified nonlinear cases particularly useful and meaningful in the engineering analysis of large deflections of thin structures. On the one hand, even when the deformations are finite, the total range of deformation can still be quite small, and the materials do not have to undergo a permanent set or failure. On the other hand, nonlinear finite deformations free us from the restrictions of the physically unrealistic situation imposed by linear infinitesimal deformations. A small nonlinearity can make a radical difference. For example, it is well known that chaotic vibrations can only take place in the presence of some nonlinearity, small as it may be, as we have discussed in the preceding chapter.

To extend from nonlinear elasticity to piezoelectricity, the most important feature includes the generalization from the strain energy density U_0 to the electric enthalpy density H, the two being related to each other by

$$H = U_0 - E_i D_i,$$

where E_i and D_i are the electric field and electric displacement, respectively. The constitutive relations are then given by

$$\sigma_{ij} = \frac{\partial H}{\partial \epsilon_{ij}}, \qquad D_i = -\frac{\partial H}{\partial E_i}, \qquad (10.1.1)$$

which are responsible for the coupling between elasticity and piezoelectricity.

In an early treatment by Toupin (1956), an attempt was made to adopt a polynomial approximation for the electric enthalpy density. Specifically, he proposed the following polynomial form of the electric enthalpy density:

$$H = H_0^A E_A + H_1^{AB} E_A E_B + H_2^{AB} \epsilon_{AB} + H_3^{ABCD} \epsilon_{AB} \epsilon_{CD} + H_4^{ABC} \epsilon_{AB} E_C$$

$$+ H_5^{ABCD} \epsilon_{AB} E_C E_D + H_6^{ABCDE} \epsilon_{AB} \epsilon_{CD} E_E + H_7^{ABCDEF} \epsilon_{AB} \epsilon_{CD} E_E E_F,$$

where a combination of his and our notations is adopted. Thus, H_0^A, H_1^{AB}, ... are Toupin's notations for material descriptors used to account for anisotropy. As he noted, the above expression can be regarded as consisting of the first several terms in a power series expansion about the natural state of an arbitrary piezoelectric medium, and it may be expected to be accurate for sufficiently small values of the strains *and* sufficiently weak fields. The terms linear in E or ϵ in the expression are not needed here and thus are neglected. By further omitting the cubic and fourth-order terms and keeping only the quadratic terms in E and/or ϵ, the above expression reduces to a form written as

$$H = \tfrac{1}{2} c_{ijkl} \epsilon_{ij} \epsilon_{kl} - e_{ijk} E_i \epsilon_{jk} - \tfrac{1}{2} e_{ij} E_i E_j. \qquad (10.1.2)$$

The right-hand side of this result is now in terms of c_{ijkl}, e_{ijk}, and e_{ij}, which are components of the elastic stiffness, piezoelectric strain constant, and dielectric permittivity, respectively.

   Combining Toupin's argument on piezoelectricity with our argument for small strains in simplified nonlinear cases of elasticity, we reach the important conclusion that, even though only a few terms are used in Eq. (10.1.2), its accuracy always can be assured in these simplified cases provided only that the electric fields are also sufficiently weak enough (Yu 1995a). Indeed, Eq. (10.1.2) represents the simplest possible form of the electric enthalpy density for these cases. Substitution of H from Eq. (10.1.2) into Eqs. (10.1.1) yields the following special form of the constitutive relations:

$$\sigma_{ij} = c_{ijkl} \epsilon_{kl} - e_{kij} E_k, \qquad D_i = e_{ijk} \epsilon_{jk} + e_{ij} E_j \qquad (10.1.3)$$

where the electric field is related to the electric potential ϕ by

$$E_i = -\phi_{,i}$$

   Equations (10.1.1) through (10.1.3) constitute the essential feature that we propose for coupled nonlinear elasticity and piezoelectricity for the simplified nonlinear cases, which is to adopt nonlinear strains in place of the linear strains that are commonly used in linear piezoelectricity.

10.2   Generalized Hamilton's Principle and Variational Equation of Motion Including Piezoelectric Effect

The generalized Hamilton's principle in Eq. (1.6.1) in nonlinear elasticity is extended to include the piezoelectric effect by letting (Yu 1995b)

$$T = \int_V \tfrac{1}{2}\rho \dot{u}_i \dot{u}_i \, dV$$

$$U = \int_V [\sigma_{ij}(\varepsilon_{ij} - \epsilon_{ij}) + H] \, dV \qquad (i, j = 1, 2, 3)$$

$$W = \int_V f_i u_i \, dV + \int_{S_p} \bar{p}_i u_i \, dS + \int_{S_u} p_i(u_i - \bar{u}_i) \, dS$$

$$\int_{S_\sigma} \bar{\sigma}\phi \, dS + \int_{S_\phi} \sigma(\phi - \bar{\phi}) \, dS.$$

$$(10.2.1)$$

The expression for T is the same as in Eqs. (1.6.2). In the expression for U, the strain energy density U_0 in Eqs. (1.6.2) now is replaced by the electric enthalpy density H, as was described in the preceding section. Finally, the expression for W in Eqs. (10.2.1) contains two more integrals than in Eqs. (1.6.2). These additional integrals involve the surface charge σ and electric potential ϕ, and S_σ and S_ϕ represent those parts of the boundary surface on which σ and ϕ are prescribed, respectively. As before, an overbar denotes a prescribed quantity.

In addition to the variations of displacements, strains, and stresses, variations of the electric field, surface charge, and electric potential also are taken independently. By substituting T, U, and W from Eqs. (10.2.1) into Eq. (1.6.1) and carrying out all of these variations, a new generalized variational equation of motion in nonlinear elasticity including the piezoelectric effect is obtained. The result is written in the form of separate components according to the various types of variations, as follows:

$$\int_{t_0}^{t_1} dt \int_V \left[\frac{1}{2} \left\{ \sigma_{ij} \left(\frac{\partial \varepsilon_{ij}}{\partial e_{mn}} + \frac{\partial \varepsilon_{ij}}{\partial \omega_{mn}} \right) \right\}_{,n} \delta_{\ell m} \right.$$

$$\left. + \frac{1}{2} \left\{ \sigma_{ij} \left(\frac{\partial \varepsilon_{ij}}{\partial e_{mn}} - \frac{\partial \varepsilon_{ij}}{\partial \omega_{mn}} \right) \right\}_{,m} \delta_{\ell n} + f_\ell - \rho \ddot{u}_\ell \right] \delta u_\ell \, dV$$

$$- \int_{t_0}^{t_1} dt \int_{S_p} \left[\frac{1}{2}\sigma_{ij} \left(\frac{\partial \varepsilon_{ij}}{\partial e_{mn}} + \frac{\partial \varepsilon_{ij}}{\partial \omega_{mn}} \right) v_n \delta_{\ell m} \right.$$

$$\left. + \frac{1}{2} \sigma_{ij} \left(\frac{\partial \varepsilon_{ij}}{\partial e_{mn}} - \frac{\partial \varepsilon_{ij}}{\partial \omega_{mn}} \right) v_m \delta_{\ell n} - \bar{p}_\ell \right] \delta u_\ell \, dS = 0 \qquad (10.2.2a)$$

$$\int_{t_0}^{t_1} dt \int_{S_u} (u_i - \bar{u}_i)\delta p_i \, dS = 0 \qquad (10.2.2b)$$

$$\int_{t_0}^{t_1} dt \int_V (\epsilon_{ij} - \varepsilon_{ij})\delta\sigma_{ij} \, dV = 0 \qquad (10.2.2c)$$

$$\int_{t_0}^{t_1} dt \int_V \left(\sigma_{ij} - \frac{\partial H}{\partial \epsilon_{ij}} \right) \delta\epsilon_{ij} \, dV = 0 \qquad (10.2.2d)$$

$$\int_{t_0}^{t_1} dt \int_V \left(D_i + \frac{\partial H}{\partial E_i} \right) \delta E_i \, dV = 0 \qquad (10.2.2e)$$

$$\int_{t_0}^{t_1} dt \int_V D_{i,i} \delta\phi \, dV - \int_{t_0}^{t_1} dt \int_{S_\sigma} (n_i D_i - \bar{\sigma})\delta\phi \, dS = 0 \qquad (10.2.2f)$$

$$\int_{t_0}^{t_1} dt \int_{S_\phi} (\phi - \bar{\phi})\delta\sigma \, dS = 0. \qquad (10.2.2g)$$

Equations (10.2.2a–g) represent the generalized variational equations of motion in coupled nonlinear elasticity and piezoelectricity. When the piezoelectric effect is suppressed, the results reduce to those in nonlinear elasticity as a special case. In fact, Eqs. (10.2.2a,b,c) have the same form regardless of whether the piezoelectric effect is included or not. Equation (10.2.2a) is also reducible to the ordinary variational equations (1.4.5), (1.4.6), (1.4.7), and eventually (1.4.8), as special cases. Similarly, as before, Eq. (10.2.2b) yields the displacement boundary conditions, and Eq. (10.2.2c) yields the relations between nonlinear strains and displacements.

Equation (10.2.2d) yields an extension of the stress–strain relations in nonlinear elasticity and, together with Eq. (10.2.2e), provides the constitutive relations in piezoelectricity that have been given in the preceding section in Eqs. (10.1.1). As mentioned earlier, when H has the form in Eq. (10.1.2), the specific constitutive relations are given by Eqs. (10.1.3).

Finally, Eqs. (10.2.2f,g) yield the field equation of electrostatics that governs the electric displacement

$$D_{i,i} = O \qquad \text{in } V, \qquad (10.2.3)$$

the boundary condition for prescribed surface charge

$$n_i D_i = \bar{\sigma} \qquad \text{on } S_\sigma, \qquad (10.2.4)$$

and the boundary condition for prescribed electric potential

$$\phi = \bar{\phi} \qquad \text{on } S_\phi. \qquad (10.2.5)$$

Equations (10.2.3) through (10.2.5) are the same as in the linear theory of piezoelectricity.

10.3   Classical Equations for Large Deflections of a Piezoelectric Plate

A classical homogeneous plate with large deflections was treated in Section 7.2. When the same displacements and rotations in Eqs. (7.2.1) and (7.2.2) are used, the *stress equations of motion* and *boundary conditions* written from the variational equation (7.2.3) remain valid even for a classical piezoelectric plate (Yu 1995a,b,c). Of course, these may also be derived from Eq. (10.2.2a). To complete the treatment for the piezoelectric plate, we make use of the other components of Eqs. (10.2.2). Integration with respect to time is now ignored.

10.3.1 Nonlinear Plate Strain–Displacement Relations

For a piezoelectric plate, these relations are also the same as those for an elastic plate, as can be determined directly from Eqs. (7.2.7). However, we shall demonstrate here the use of the component equation (10.2.2c). Carrying out integration with respect to z, we find

$$
\int_V (\epsilon_{ij} - \varepsilon_{ij})\delta\sigma_{ij}\, dV
$$

$$
= \int\int_A \int_{-h}^{h} \left\{ \left[\epsilon_{xx}^{(0)} + z\epsilon_{xx}^{(1)} - \frac{\partial u}{\partial x} + z\frac{\partial^2 w}{\partial x^2} - \frac{1}{2}\left(\frac{\partial w}{\partial x}\right)^2 \right]\delta\sigma_{xx} \right.
$$

$$
+ \left[\epsilon_{yy}^{(0)} + z\epsilon_{yy}^{(1)} - \frac{\partial v}{\partial y} + z\frac{\partial^2 w}{\partial y^2} - \frac{1}{2}\left(\frac{\partial w}{\partial y}\right)^2 \right]\delta\sigma_{yy}
$$

$$
+ \left[\epsilon_{xy}^{(0)} + z\epsilon_{xy}^{(1)} - \frac{1}{2}\left(\frac{\partial u}{\partial y} + \frac{\partial v}{\partial x}\right) + z\frac{\partial^2 w}{\partial x \partial y} \right.
$$

$$
\left. - \frac{1}{2}\frac{\partial w}{\partial x}\frac{\partial w}{\partial y} \right]\delta\sigma_{xy} + \left[\epsilon_{zx}^{(0)}\right]\delta\sigma_{zx} + \left[\epsilon_{zy}^{(0)}\right]\delta\sigma_{zy}
$$

$$
\left. + \left[\epsilon_{zz}^{(0)} + z\epsilon_{zz}^{(1)}\right]\delta\sigma_{zz} \right\} dx\, dy\, dz \tag{10.3.1}
$$

$$
= \int\int_A \left\{ \left[\epsilon_{xx}^{(0)} - \frac{\partial u}{\partial x} - \frac{1}{2}\left(\frac{\partial w}{\partial x}\right)^2 \right]\delta N_x + \left[\epsilon_{xx}^{(1)} + \frac{\partial^2 w}{\partial x^2}\right]\delta M_x \right.
$$

$$
+ \left[\epsilon_{yy}^{(0)} - \frac{\partial v}{\partial y} - \frac{1}{2}\left(\frac{\partial w}{\partial y}\right)^2 \right]\delta N_y + \left[\epsilon_{yy}^{(1)} + \frac{\partial^2 w}{\partial y^2}\right]\delta M_y
$$

$$
+ \left[\epsilon_{xy}^{(0)} - \frac{1}{2}\left(\frac{\partial u}{\partial y} + \frac{\partial v}{\partial x}\right) - \frac{1}{2}\frac{\partial w}{\partial x}\frac{\partial w}{\partial y} \right]\delta N_{xy}
$$

$$
+ \left[\epsilon_{xy}^{(1)} + \frac{\partial^2 w}{\partial x \partial y}\right]\delta M_{xy} + \left[\epsilon_{zx}^{(0)}\right]\delta Q_x + \left[\epsilon_{zy}^{(0)}\right]\delta Q_y
$$

$$
\left. + \left[\epsilon_{zz}^{(0)}\right]\delta\sigma_{zz}^{(0)} + \left[\epsilon_{zz}^{(1)}\right]\delta\sigma_{zz}^{(1)} \right\} dx\, dy = 0,
$$

where a dual system of notations for the plate stresses is adopted, namely,

$$
N_x = \sigma_{xx}^{(0)} = \int_{-h}^{h}\sigma_{xx}\, dz, \qquad N_y = \sigma_{yy}^{(0)} = \ldots, \qquad N_{xy} = \sigma_{xy}^{(0)} = \ldots
$$

$$
M_x = \sigma_{xx}^{(1)} = \int_{-h}^{h}\sigma_{xx}z\, dz, \qquad M_y = \sigma_{yy}^{(1)} = \ldots, \qquad M_{xy} = \sigma_{xy}^{(1)} = \ldots .
$$

Relations between the plate strains and plate displacements are obtained from Eq. (10.3.1) by setting equal to zero the coefficients of δN_x, δM_x, $\ldots$. The transverse shear plate strains $\epsilon_{zx}^{(0)}$ and $\epsilon_{zy}^{(0)}$ vanish, but the transverse shear forces Q_x

and Q_y need not. On the other hand, the transverse normal plate strains $\epsilon_{zz}^{(0)}$ and $\epsilon_{zz}^{(1)}$ are allowed to develop freely, although the transverse normal plate stresses $\sigma_{zz}^{(0)}$ and $\sigma_{zz}^{(1)}$ do not appear in the stress equations of motion and thus can be set equal to zero. These are important characteristics associated with a classical plate.

10.3.2 Constitutive Relations

The constitutive relations for a linear piezoelectric plate have been discussed in detail in the works of Mindlin, Tiersten, and Lee et al. that were cited earlier. Some of these were derived from the three-dimensional linear piezoelectricity and are now extended to cover nonlinear strains as indicated in Eqs. (10.1.1) through (10.1.3). Equations (10.1.3) also have appeared in the form of Eqs. (10.2.2d,e) as components of the three-dimensional generalized variational equation, which are used here to derive the constitutive relations for a piezoelectric plate with large deflections.

We start with Eq. (10.2.2d), now written in the form

$$\int\int_A\int_{-h}^{h} \left(\sigma_{ij} - c_{ijkl}\,\epsilon_{kl} + e_{kij}E_k\right)\delta\epsilon_{ij}\,dx\,dy\,dz = 0. \tag{10.3.2}$$

Incorporating the first-order approximations

$$\epsilon_{kl} = \epsilon_{kl}^{(0)} + z\epsilon_{kl}^{(1)} \qquad (k, \ell = x, y, z)$$

$$E_k = E_k^{(0)} + z\,E_k^{(1)} \qquad (k = x, y, z) \tag{10.3.3}$$

and integrating with respect to z, we readily find

$$\int\int_A \left\{ \left[\sigma_{ij}^{(0)} - 2h\left(c_{ijkl}\epsilon_{kl}^{(0)} - e_{kij}E_k^{(0)}\right)\right]\delta\epsilon_{ij}^{(0)} \right.$$
$$\left. + \left[\sigma_{ij}^{(1)} - \frac{2}{3}h^3\left(c_{ijkl}\epsilon_{kl}^{(1)} - e_{kij}E_k^{(1)}\right)\right]\delta\epsilon_{ij}^{(1)} \right\}dx\,dy = 0. \tag{10.3.4}$$

In component form, the integrand in Eq. (10.3.4) consists of six pairs of relations corresponding to the six pairs of variations of the plate strains. As was mentioned above in connection with Eq. (10.3.1), the transverse shear plate strains $\epsilon_{zy}^{(0)}$, $\epsilon_{zx}^{(0)}$, $\epsilon_{zy}^{(1)}$, and $\epsilon_{zx}^{(1)}$ are zero in the classical plate theory. The two corresponding pairs of relations therefore drop out. A third pair that also is associated with the z-direction has the form

$$\left[\sigma_{zz}^{(0)} - 2h\left(c_{zzkl}\,\epsilon_{kl}^{(0)} - e_{kzz}E_k^{(0)}\right)\right]\delta\epsilon_{zz}^{(0)}$$
$$+ \left[\sigma_{zz}^{(1)} - \tfrac{2}{3}h^3\left(c_{zzkl}\,\epsilon_{kl}^{(1)} - e_{kzz}E_k^{(1)}\right)\right]\delta\epsilon_{zz}^{(1)},$$

in which the transverse plate normal stresses $\sigma_{zz}^{(0)}$ and $\sigma_{zz}^{(1)}$ can be set equal to zero, as was mentioned earlier. By equating the coefficients of $\delta\epsilon_{zz}^{(0)}$ and $\delta\epsilon_{zz}^{(1)}$ to zero,

$\epsilon_{zz}^{(0)}$ and $\epsilon_{zz}^{(1)}$ then can be solved in terms of the other plate strains and plate electric fields. Indeed, when the results are substituted into the remaining three pairs of relations, $\epsilon_{zz}^{(0)}$ and $\epsilon_{zz}^{(1)}$ can be eliminated. Equation (10.3.4) then reduces to the final form

$$\int\int_A \left\{ \left[\sigma_{ij}^{(0)} - 2h \left(\overset{*}{c}_{ijkl}\, \epsilon_{kl}^{(0)} - \overset{*}{e}_{kij}\, E_k^{(0)} \right) \right] \delta\epsilon_{ij}^{(0)} \right.$$

$$\left. + \left[\sigma_{ij}^{(1)} - \tfrac{2}{3}h^3 \left(\overset{*}{c}_{ijkl}\, \epsilon_{kl}^{(1)} - \overset{*}{e}_{kij}\, E_k^{(1)} \right) \right] \delta\epsilon_{ij}^{(1)} \right\} dx\, dy = 0,$$

(10.3.5)

where

$$\overset{*}{c}_{ijkl} = c_{ijkl} - c_{zzkl}\, \frac{c_{ijzz}}{c_{zzzz}}$$

$$\overset{*}{e}_{kij} = e_{kij} - e_{kzz}\, \frac{c_{ijzz}}{c_{zzzz}}.$$

Equation (10.3.5) now yields only three pairs of plate stress–strain relations, with subscripts i and j equal to x or y, but no longer z. These include $\sigma_{xx}^{(0)}$, $\sigma_{xx}^{(1)}$, $\sigma_{yy}^{(0)}$, $\sigma_{yy}^{(1)}$, $\sigma_{xy}^{(0)}$, and $\sigma_{xy}^{(1)}$ or N_x, M_x, N_y, M_y, N_{xy}, and M_{xy}.

We next consider Eq. (10.2.2e) similarly, which is written as

$$\int\int_A \int_{-h}^{h} \left(D_i - e_{ijk}\,\epsilon_{jk} - e_{ij}E_j \right) \delta E_i\, dx\, dy\, dz = 0. \qquad (10.3.6)$$

Substitution of Eqs. (10.3.3) into (10.3.6) and integration with respect to z yield

$$\int\int_A \left\{ \left[D_i^{(0)} - 2h \left(e_{ijk}\, \epsilon_{jk}^{(0)} + e_{ij}\, E_j^{(0)} \right) \right] \delta E_j^{(0)} \right.$$

$$\left. + \left[D_i^{(1)} - \tfrac{2}{3}h^3 \left(e_{ijk}\, \epsilon_{jk}^{(1)} + e_{ij}\, E_j^{(1)} \right) \right] \delta E_j^{(1)} \right\} dx\, dy = 0.$$

As in the situation with the plate stresses, this is modified by replacing e_{ijk} by $\overset{*}{e}_{ijk}$ so that the result becomes

$$\int\int_A \left\{ \left[D_i^{(0)} - 2h \left(\overset{*}{e}_{ijk}\, \epsilon_{jk}^{(0)} + e_{ij}\, E_j^{(0)} \right) \right] \delta E_j^{(0)} \right.$$

$$\left. + \left[D_i^{(1)} - \tfrac{2}{3}h^3 \left(\overset{*}{e}_{ijk}\, \epsilon_{jk}^{(1)} + e_{ij}\, E_j^{(1)} \right) \right] \delta E_j^{(1)} \right\} dx\, dy = 0,$$

from which the relations for the plate electric displacements can be written immediately.

The plate electric enthalpy density is now

$$H^{(0)} = \int_{-h}^{h} H\, dz = h \left(\overset{*}{c}_{ijkl}\epsilon_{ij}^{(0)}\epsilon_{kl}^{(0)} - 2\overset{*}{e}_{kij}E_k^{(0)}\epsilon_{ij}^{(0)} - e_{ij}E_i^{(0)}E_j^{(0)} \right)$$

$$+ \tfrac{1}{3}h^3 \left(\overset{*}{c}_{ijkl}\epsilon_{ij}^{(1)}\epsilon_{kl}^{(1)} - 2\overset{*}{e}_{kij}E_k^{(1)}\epsilon_{ij}^{(1)} - e_{ij}E_i^{(1)}E_j^{(1)} \right),$$

from which

$$\sigma_{ij}^{(0)} = \frac{\partial H^{(0)}}{\partial \epsilon_{ij}^{(0)}}, \qquad \sigma_{ij}^{(1)} = \frac{\partial H^{(0)}}{\partial \epsilon_{ij}^{(1)}}$$

$$D_i^{(0)} = -\frac{\partial H^{(0)}}{\partial E_i^{(0)}}, \qquad D_i^{(1)} = -\frac{\partial H^{(0)}}{\partial E_i^{(1)}}.$$

10.3.3 Field Equations and Boundary Conditions in Electrostatics for Plate

Field equations of electrostatics are derived from Eq. (10.2.2f) in a similar manner as stress equations of motion are derived from Eq. (10.2.2a). Each of these three-dimensional variational equations contains a volume integral and a surface integral, with the former applied to the volume of the plate and the latter applied to the boundary planes $z = \pm h$ and to the cylindrical boundary surface of the plate. In deriving the stress equations, we allowed the prescription of face tractions on $z = \pm h$. In deriving the equations of electrostatics, we similarly allow the prescription of face charges on $z = \pm h$.

Introducing the first-order approximation of the electric potential

$$\phi = \phi^{(0)} + z\,\phi^{(1)}, \tag{10.3.7}$$

we find from Eq. (10.2.2f)

$$\int \int_A \left\{ \left[\frac{\partial D_x^{(0)}}{\partial x} + \frac{\partial D_y^{(0)}}{\partial y} + \Sigma^{(0)} \right] \delta\phi^{(0)} \right.$$
$$\left. + \left[\frac{\partial D_x^{(1)}}{\partial x} + \frac{\partial D_y^{(1)}}{\partial y} - D_z^{(0)} + \Sigma^{(1)} \right] \delta\phi^{(1)} \right\} dx\,dy$$
$$- \int_{C_\sigma} \left\{ \left[D_n^{(0)} - \sigma^{(0)} \right] \delta\phi^{(0)} + \left[D_n^{(1)} - \sigma^{(1)} \right] \delta\phi^{(1)} \right\} ds = 0, \tag{10.3.8}$$

where C_σ is that part of the boundary curve C on which the edge charges are prescribed,

$$\left(D_i^{(0)}, D_i^{(1)} \right) = \int_{-h}^{h} (D_i, D_i z)\,dz, \qquad \left(\sigma^{(0)}, \sigma^{(1)} \right) = \int_{-h}^{h} (\sigma, \sigma z)\,dz$$

are the plate electric displacements and plate edge charges, respectively, and

$$\Sigma^{(0)} = (\sigma)_h + (\sigma)_{-h}, \qquad \Sigma^{(1)} = h[(\sigma)_h - (\sigma)_{-h}]$$

are the plate face charges. Equation (10.3.8) represents another component of the two-dimensional variational equation for the piezoelectric plate, from which the field equations of electrostatics and edge conditions for prescribed electric charge can be written immediately.

An alternative set of edge conditions for prescribed electric potential is derived from Eq. (10.2.2g). By introducing the same approximation in Eq. (10.3.7), Eq. (10.2.2g) becomes

$$\int_{C_\phi} \left\{ \left[\phi^{(0)} - \bar{\phi}^{(0)} \right] \delta \sigma^{(0)} + \left[\phi^{(1)} - \bar{\phi}^{(1)} \right] \delta \sigma^{(1)} \right\} ds = 0, \qquad (10.3.9)$$

from which edge potential conditions can be written similarly.

Prescription of face potentials is an important practical consideration for piezoelectric layers used in electromechanical transducers. With adoption of the first order approximation of the electric potential in Eq. (10.3.7), prescription of uniform face potentials on the boundary planes $z = \pm h$ can be accommodated without difficulty, as shown by Mindlin (1974). Similarly, for higher order approximations based on expansions in series of trigonometric functions, no difficulty was encountered by Bugdayci and Bogy (1981) by using Eq. (10.3.7) in conjunction with a sine series. When a cosine series was used by Lee et al. (1987), however, a relation connecting face charge to face potential had to be introduced to accommodate the prescription of face potentials.

10.4   Refined Equations for Large Deflections of a Piezoelectric Plate

Consider the same piezoelectric plate as discussed in the preceding section. The three-dimensional displacements now are taken in the form of the following first-order approximation:

$$u_x = u + z\psi, \qquad u_y = v + z\varphi, \qquad u_z = w + z\beta, \qquad (10.4.1)$$

where the two-dimensional plate displacements u, v, w, ψ, ϕ, and β are independent of z. Since the displacements in Eqs. (10.4.1) are the same as those used by Mindlin (1984) in the final version of his treatment of the linear piezoelectric plate, the results obtained here will be the counterpart of his for large deflections. As he noted, these accommodate the thickness-shear, thickness-twist, and thickness-stretch modes of vibration in an infinite linear plate. The rotations are taken in the same form as before:

$$\omega_{yz} = \frac{\partial w}{\partial y}, \qquad \omega_{zx} = -\frac{\partial w}{\partial x}, \qquad \omega_{xy} = 0, \qquad (10.4.2)$$

which are also independent of z.

10.4.1   Stress Equations of Motion and Boundary Conditions for Plate

By substituting Eqs. (10.4.1) and (10.4.2) into the three-dimensional variational equation (10.2.2a) or (1.4.8) and carrying out integration with respect to z over the plate thickness, there results

$$
\int\int_A \left\{ \left[\frac{\partial N_x}{\partial x} + \frac{\partial N_{xy}}{\partial y} + P_x^{(0)} + f_x^{(0)} - 2\rho h \ddot{u} \right] \delta u \right.
$$

$$
+ \left[\frac{\partial N_{xy}}{\partial x} + \frac{\partial N_y}{\partial y} + P_x^{(0)} + f_y^{(0)} - 2\rho h \ddot{v} \right] \delta v
$$

$$
+ \left[\frac{\partial M_x}{\partial x} + \frac{\partial M_{xy}}{\partial y} - Q_x + P_x^{(1)} + f_x^{(1)} - \frac{2}{3}\rho h^3 \ddot{\psi} \right] \delta\psi
$$

$$
+ \left[\frac{\partial M_{xy}}{\partial x} + \frac{\partial M_y}{\partial y} - Q_y + P_y^{(1)} + f_y^{(1)} - \frac{2}{3}\rho h^3 \ddot{\phi} \right] \delta\phi
$$

$$
+ \left[\frac{\partial}{\partial x}\left(Q_x + N_x \frac{\partial w}{\partial x} + N_{xy} \frac{\partial w}{\partial y} \right) + \frac{\partial}{\partial y}\left(Q_y + N_{xy} \frac{\partial w}{\partial x} + N_y \frac{\partial w}{\partial y} \right) \right.
$$

$$
\left. + P_z^{(0)} + f_z^{(0)} - 2\rho h \ddot{w} \right] \delta w
$$

$$
+ \left[\frac{\partial}{\partial x}\left(S_x + M_x \frac{\partial w}{\partial x} + M_{xy} \frac{\partial w}{\partial y} \right) + \frac{\partial}{\partial y}\left(S_y + M_{xy} \frac{\partial w}{\partial x} + M_y \frac{\partial w}{\partial y} \right) \right.
$$

$$
\left. \left. - \left(Q_z + Q_x \frac{\partial w}{\partial x} + Q_y \frac{\partial w}{\partial y} \right) + P_z^{(1)} + f_z^{(1)} - \frac{2}{3}\rho h^3 \ddot{\beta} \right] \delta\beta \right\} dx\,dy
$$

$$
\oint_{C_P} \left\{ \left[N_n - p_n^{(0)} \right] \delta u_n + \left[N_{ns} - p_s^{(0)} \right] \delta u_s \right.
$$

$$
+ \left[M_n - p_n^{(1)} \right] \delta\psi_n + \left[M_{ns} - p_s^{(1)} \right] \delta\psi_s
$$

$$
+ \left[Q_n + N_n \frac{\partial w}{\partial n} + N_{ns} \frac{\partial w}{\partial s} - p_z^{(0)} \right] \delta w
$$

$$
\left. + \left[S_n + M_n \frac{\partial w}{\partial n} + M_{ns} \frac{\partial w}{\partial s} - p_z^{(1)} \right] \delta\beta \right\} ds = 0, \qquad (10.4.3)
$$

where C_p is that part of the contour C on which tractions are prescribed, n and s refer to coordinates normal and tangential to the contour, and other notations are

$$
(N_x, N_y, N_{xy}, M_x, M_y, M_{xy}) = \int_{-h}^{h} (\sigma_{xx}, \sigma_{yy}, \sigma_{xy}, \sigma_{xx}z, \sigma_{yy}z, \sigma_{xy}z)\, dz
$$

$$
(Q_x, Q_y, Q_z, S_x, S_y) = \int_{-h}^{h} (\sigma_{zx}, \sigma_{zy}, \sigma_{zz}, \sigma_{zx}z, \sigma_{zy}z)\, dz
$$

$$P_x^{(0)} = (p_x)\,h + (p_x)_{-h}\,, \qquad P_y^{(0)} = \ldots, \qquad P_z^{(0)} = \ldots$$

$$P_x^{(1)} = h\left[(p_x)_h - (p_x)_{-h}\right], \qquad P_y^{(1)} = \ldots, \qquad P_z^{(1)} = \ldots$$

$$\left(f_x^{(0)}, f_x^{(1)}, f_y^{(0)}, f_y^{(1)}, f_z^{(0)}, f_z^{(1)}\right) = \int_{-h}^{h} \left(f_x, f_x z, f_y, f_y z, f_z, f_z z\right)\,dz$$

$$\left(p_n^{(0)}, p_s^{(0)}, p_z^{(0)}, p_n^{(1)}, p_s^{(1)}, p_z^{(1)}\right) = \int_{-h}^{h} \left(p_n, p_s, p_z, p_n z, p_s z, p_z z\right)\,dz.$$

$$(10.4.4)$$

Equation (10.4.3) is a pseudo-variational equation of the piezoelectric plate from which the stress equations of motion and traction boundary conditions in the two-dimensional plate theory can be written readily.

The displacement boundary conditions are similarly derived from Eq. (10.2.2b). By integrating over the thickness of the plate, the three-dimensional equation reduces to a two-dimensional one, as follows:

$$\oint_{C_u} \left\{ \left[u_n - \bar{u}_n\right]\delta p_n^{(0)} + \left[\psi_n - \bar{\psi}_n\right]\delta p_n^{(1)} \right.$$

$$+ \left[u_s - \bar{u}_s\right]\delta p_s^{(0)} + \left[\psi_s - \bar{\psi}_s\right]\delta p_s^{(1)} \qquad (10.4.5)$$

$$\left. + \left[w - \bar{w}\right]\delta p_z^{(0)} + \left[\beta - \bar{\beta}\right]\delta p_z^{(1)} \right\}\,ds = 0,$$

where C_u is that part of C on which displacements are prescribed. By equating the coefficients of the variations in Eq. (10.4.5) to 0, the displacement boundary conditions in the two-dimensional plate theory are obtained.

10.4.2 Nonlinear Plate Strain–Displacement Relations

Corresponding to the displacements in Eqs. (10.4.1) and the rotations in Eqs. (10.4.2), the nonlinear strains in the plate theory have the form

$$\epsilon_{xx} = \frac{\partial u_x}{\partial x} + \frac{1}{2}\omega_{zx}^2 = \frac{\partial u}{\partial x} + z\frac{\partial \psi}{\partial x} + \frac{1}{2}\left(\frac{\partial w}{\partial x}\right)^2$$

$$\epsilon_{yy} = \frac{\partial u_y}{\partial y} + \frac{1}{2}\omega_{zy}^2 = \frac{\partial v}{\partial y} + z\frac{\partial \phi}{\partial y} + \frac{1}{2}\left(\frac{\partial w}{\partial y}\right)^2$$

$$\epsilon_{zz} = \frac{\partial u_z}{\partial z} + \frac{1}{2}\omega_{xy}^2 = \beta$$

$$\epsilon_{xy} = \frac{1}{2}\left(\frac{\partial u_x}{\partial y} + \frac{\partial u_y}{\partial x}\right) - \omega_{zx}\omega_{zy} \qquad (10.4.6)$$

$$= \frac{1}{2}\left(\frac{\partial u}{\partial y} + \frac{\partial v}{\partial x} + z\frac{\partial \psi}{\partial y} + z\frac{\partial \phi}{\partial x}\right) + \frac{\partial w}{\partial x}\frac{\partial w}{\partial y}$$

$$\epsilon_{zx} = \frac{1}{2}\left(\frac{\partial u_x}{\partial z} + \frac{\partial u_z}{\partial x}\right) - \omega_{xy}\omega_{zy} = \frac{1}{2}\left(\psi + \frac{\partial w}{\partial x} + z\frac{\partial \beta}{\partial x}\right)$$

$$\epsilon_{zy} = \frac{1}{2}\left(\frac{\partial u_y}{\partial z} + \frac{\partial u_z}{\partial y}\right) - \omega_{xy}\omega_{zx} = \frac{1}{2}\left(\phi + \frac{\partial w}{\partial y} + z\frac{\partial \beta}{\partial y}\right).$$

These also can be written in the form of the first-order approximations

$$\epsilon_{xx} = \epsilon_{xx}^{(0)} + z\epsilon_{xx}^{(1)}, \qquad \epsilon_{yy} = \epsilon_{yy}^{(0)} + z\epsilon_{yy}^{(1)}, \qquad \epsilon_{zz} = \epsilon_{zz}^{(0)} + z\epsilon_{zz}^{(1)}$$
$$\epsilon_{xy} = \epsilon_{xy}^{(0)} + z\epsilon_{xy}^{(1)}, \qquad \epsilon_{zx} = \epsilon_{zx}^{(0)} + z\epsilon_{zx}^{(1)}, \qquad \epsilon_{zy} = \epsilon_{zy}^{(0)} + z\epsilon_{zy}^{(1)}, \tag{10.4.7}$$

where the superscripts 0 and 1 within parentheses designate as before the plate strains of the zeroth and first orders in z, respectively. While relations between the plate strains and plate displacements can be determined directly from Eqs. (10.4.6) and (10.4.7), they also can be derived from the variational Eq. (10.2.2c) by integration over the thickness of the plate in the same manner as the similar relations for the classical case were derived in the preceding section. The results are

$$\int_A \int \left\{ \left[\epsilon_{xx}^{(0)} - \frac{\partial u}{\partial x} - \frac{1}{2}\left(\frac{\partial w}{\partial x}\right)^2 \right] \delta N_x + \left[\epsilon_{xx}^{(1)} - \frac{\partial \psi}{\partial x} \right] \delta M_x \right.$$
$$+ \left[\epsilon_{yy}^{(0)} - \frac{\partial v}{\partial y} - \frac{1}{2}\left(\frac{\partial w}{\partial y}\right)^2 \right] \delta N_y + \left[\epsilon_{yy}^{(1)} - \frac{\partial \phi}{\partial y} \right] \delta M_y$$
$$+ \left[\epsilon_{zz}^{(0)} - \beta \right] \delta Q_z + \left[\epsilon_{zz}^{(1)} \right] \delta \sigma_{zz}^{(1)}$$
$$+ \left[\epsilon_{xy}^{(0)} - \frac{1}{2}\left(\frac{\partial u}{\partial y} + \frac{\partial v}{\partial x}\right) - \frac{\partial w}{\partial x}\frac{\partial w}{\partial y} \right] \delta N_{xy}$$
$$+ \left[\epsilon_{xy}^{(1)} - \frac{1}{2}\left(\frac{\partial \psi}{\partial y} + \frac{\partial \phi}{\partial x}\right) \right] \delta M_{xy}$$
$$+ \left[\epsilon_{zx}^{(0)} - \frac{1}{2}\left(\psi + \frac{\partial w}{\partial x}\right) \right] \delta Q_x + \left[\epsilon_{zx}^{(1)} - \frac{1}{2}\frac{\partial \beta}{\partial x} \right] \delta S_x$$
$$\left. + \left[\epsilon_{zy}^{(0)} - \frac{1}{2}\left(\phi + \frac{\partial w}{\partial y}\right) \right] \delta Q_y + \left[\epsilon_{zy}^{(1)} - \frac{1}{2}\frac{\partial \beta}{\partial y} \right] \delta S_y \right\} dx\,dy = 0. \tag{10.4.8}$$

The plate strain–displacement relations then are obtained by setting equal to 0 the coefficients of the variations of the plate stresses. In particular, setting the coefficient of $\delta\sigma_{zz}^{(1)}$ equal to zero yields $\epsilon_{zz}^{(1)} = 0$, which is consistent with Eqs. (10.4.6) and (10.4.7).

10.4.3 Constitutive Relations

To develop the two-dimensional constitutive relations for the piezoelectric plate, we start with Eq. (10.2.2d), now written in the same form as in Eq. (10.3.2). By incorporating the same first-order approximations as given in Eqs. (10.3.3), Eq. (10.3.4) is again obtained. We then derive from the latter equation

$$\sigma_{ij}^{(0)} = 2h\left(c_{ijkl}\epsilon_{kl}^{(0)} - e_{kij}E_k^{(0)}\right) \tag{10.4.9}$$

$$\sigma_{ij}^{(1)} = \tfrac{2}{3}h^3 \left(c_{ijkl}\epsilon_{kl}^{(1)} - e_{kij}E_k^{(1)}\right).$$ (10.4.10)

Relations for $\sigma_{ij}^{(0)}$ in Eq. (10.4.9) are now already in their final form, in contrast to the classical case. Those for $\sigma_{ij}^{(1)}$ in Eq. (10.4.10) will be modified further to accommodate the situation that $\sigma_{zz}^{(1)}$ does not actually appear in the stress equations of motion and traction boundary conditions. By setting $\sigma_{zz}^{(1)}$ equal to zero and having $\epsilon_{zz}^{(1)}$ eliminated, Eq. (10.4.10) becomes

$$\sigma_{ij}^{(1)} = \tfrac{2}{3}h^3 \left(c_{ijkl}^*\epsilon_{kl}^{(1)} - e_{kij}^*E_k^{(1)}\right),$$ (10.4.11)

where, as before,

$$c_{ijkl}^* = c_{ijkl} - \frac{c_{zzkl}c_{ijzz}}{c_{zzzz}}$$

$$e_{kij}^* = e_{kij} - \frac{e_{kzz}c_{ijzz}}{c_{zzzz}}.$$

The results for $\sigma_{ij}^{(1)}$ remain to have the same form as in the classical case.

Next we consider Eq. (10.2.2e) similarly, which is written in the same form as Eq. (10.3.6). Substitution of the same first-order approximations as given in Eq. (10.3.3) and integration with respect to z yield the same results as in the classical case:

$$D_i^{(0)} = 2h \left(e_{ijkl}\epsilon_{jk}^{(0)} + e_{ij}E_j^{(0)}\right)$$ (10.4.12)

$$D_i^{(1)} = \tfrac{2}{3}h^3 \left(e_{ijk}\epsilon_{jk}^{(1)} + e_{ij}E_j^{(1)}\right).$$ (10.4.13)

However, similar to the relations for the plate stresses, those for $D_i^{(0)}$ in Eq. (10.4.12) are already in their final form. Only those for $D_i^{(1)}$ in Eq. (10.4.13) need be modified by replacing e_{ijk} by e_{ijk}^* so that they become

$$D_i^{(1)} = \tfrac{2}{3}h^3 \left(e_{ijk}^*\epsilon_{jk}^{(1)} + e_{ij}E_j^{(1)}\right).$$

The plate electric enthalpy density is then

$$H^{(0)} = \int_{-h}^{h} H\, dz = h \left(c_{ijkl}\epsilon_{ij}^{(0)}\epsilon_{kl}^{(0)} - 2e_{kij}E_k^{(0)}\epsilon_{ij}^{(0)} - e_{ij}E_i^{(0)}E_j^{(0)}\right)$$
$$+ \frac{1}{3}h^3 \left(c_{ijkl}^*\epsilon_{ij}^{(1)}\epsilon_{kl}^{(1)} - 2e_{kij}^*E_k^{(1)}\epsilon_{ij}^{(1)} - e_{ij}E_i^{(1)}E_j^{(1)}\right),$$

from which

$$\sigma_{ij}^{(0)} = \frac{\partial H^{(0)}}{\partial \epsilon_{ij}^{(0)}}$$

$$\sigma_{ij}^{(1)} = \frac{\partial H^{(0)}}{\partial \epsilon_{ij}^{(1)}}$$

$$D_i^{(0)} = -\frac{\partial H^{(0)}}{\partial E_i^{(0)}}$$

$$D_i^{(1)} = -\frac{\partial H^{(0)}}{\partial E_i^{(1)}}.$$

Three correction factors were adopted by Mindlin (1984) to make the three simple thickness frequencies obtained from the linear piezoelectric plate equations the same as those obtained from the three-dimensional equations. However, it should be noted that the correction factors are not the same as the shear factor he introduced earlier (Mindlin 1951a,b). For the electroded SC-cut quartz plates that he calculated (Mindlin 1984), all three correction factors came out to be very close to the value of 1.005. Since a value of the correction factor equal to 1 is equivalent to a shear factor equal to $\pi^2/12$, the results apparently indicated that the effect of the two surface electrode layers was indeed quite small.

10.4.4 Field Equations and Boundary Conditions in Electrostatics for Plate

These are derived from Eqs. (10.2.2f,g) in the same way as the classical piezoelectric plate was treated in the preceding section. By introducing the same approximation for ϕ given in Eq. (10.3.7), these yield the same results for the plate as given in Eqs. (10.3.8) and (10.3.9). The results are also the same as those for linear piezoelectricity.

10.5 Final Remarks on the Variational Equations of Motion

In the last section in Chapter 7, we made some remarks on the use of the variational equations of motion. We conclude this chapter, and in fact this book, by reinforcing some of the earlier remarks in light of the extension to piezoelectricity.

We have noted that, as demonstrated in this book, the three-dimensional variational equations of motion can be used to derive a variety of two-dimensional plate equations. We again would like to call attention to the fact that the same variational equations of motion are also very powerful tools for carrying out approximations (Yu 1974). Specifically, we have suggested the creation of meaningful and effective connections between the variational equations on the one hand and the approximate numerical methods on the other hand. The latter may include finite element, boundary element, and other numerical methods. Now that the piezoelectric effect has further been taken into account in not only linear but also nonlinear cases, we would like to propose a further extension by encouraging those interested to go on with developing the numerical analysis of piezoelectric layers through the use of the variational equations of motion discussed in this chapter. Moreover, the piezoelectric layers can even be combined with the use of homogeneous, sandwich, or laminated composite layers, as treated in previous chapters.

References

Baily, T. and J.E. Hubbard. (1987) Distributed Piezoelectric Polymer Active Vibration Control of a Cantilever Beam. *Journal of Guidance, Control and Dynamics*, Vol. 8, pp. 605–611.

Bugdayci, N. and D.B. Bogy. (1981) A Two-Dimensional Theory for Piezoelectric Layers Used in Electro-Mechanical Transducers—I. Derivation and II. Applications. *International Journal of Solids and Structures*, Vol. 17, pp. 1159–1178, 1179–1202.

Janas, V.E. and A. Safari. (1995) Overview of Fine-Scale Piezoelectric Ceramic/ Polymer Composite Processing. *Journal of the American Ceramic Society*, Vol. 78, pp. 2945–2955.

Lee, C.K. and F.C. Moon. (1990) Modal Sensors/Actuators. *Journal of Applied Mechanics*, Vol. 57, pp. 434–441.

Lee, P.C.Y., S. Syngellakis, and J.P. Hou. (1987) A Two-Dimensional Theory for High-Frequency Vibrations of Piezoelectric Crystal Plates With or Without Electrodes. *Journal of Applied Physics*, Vol. 61, pp. 1249–1262.

Mindlin, R.D. (1951a) Influence of Rotatory Inertia and Shear on Flexural Motions of Isotropic, Elastic Plates. *Journal of Applied Mechanics*, Vol. 18, pp. 31–38.

Mindlin, R.D. (1951b) Thickness-Shear and Flexural Vibrations of Crystal Plates. *Journal of Applied Physics*, Vol. 22, pp. 316–323.

Mindlin, R.D. (1952) Forced Thickness-Shear and Flexural Vibrations of Piezoelectric Crystal Plates. *Journal of Applied Physics*, Vol. 23, pp. 83–88.

Mindlin, R.D. (1972) High Frequency Vibrations of Piezoelectric Crystal Plates. *International Journal of Solids and Structures*, Vol. 8, pp. 887–906.

Mindlin, R.D. (1974) Coupled Piezoelectric Vibrations of Quartz Plates. *International Journal of Solids and Structures*, Vol. 10, pp. 453–459.

Mindlin, R.D. (1984) Frequencies of Piezoelectrically Forced Vibrations of Electroded, Doubly Rotated, Quartz Plates. *International Journal of Solids and Structures*, Vol. 20, pp. 141–157.

Newnham, R.E. and G.R. Ruschau. (1991) Smart Electroceramics. *Journal of the American Ceramic Society*, Vol. 74, pp. 463–480.

Tiersten, H.F. and R.D. Mindlin. (1962) Forced Vibrations of Piezoelectric Crystal Plates. *Quarterly of Applied Mathematics*, Vol. 20, pp. 107–119.

Tiersten, H.F. (1969) *Linear Piezoelectric Plate Vibrations*. Plenum Press, New York.

Toupin, R.A. (1956) The Elastic Dielectric. *Journal of Rational Mechanics and Analysis*, Vol. 5, pp. 849–916.

Tzou, H.S. (1991) Distributed Modal Identification and Vibration Control of Continua: Theory and Applications, *Journal of Dynamic Systems, Measurements, and Control*, Vol. 113, pp. 494–499.

Yu, Y.Y. (1968) Stability of Nonlinear Attitude Control Systems, Including Particularly Effect of Large Deflection of Space Vehicles. In: *Proceedings of the 19th Congress of International Astronautical Federation*, pp. 341–360.

Yu, Y.Y. (1974) Application of Variational and Galerkin Equations to Linear and Nonlinear Finite Element Analysis. *Proceedings of the 25th Congress of International Astronautical Federation*, Amsterdam.

Yu, Y.Y. (1992) Equations for Large Deflections of Elastic and Piezoelectric Plates and Shallow Shells, Including Sandwiches and Laminated Composites, with Applications to Vibrations, Chaos, and Acoustic Radiation. Presented at the XIIIth International Congress of Theoretical and Applied Mechanics, Haifa, Israel.

Yu, Y.Y. (1995a) On Small Strains and Large Rotations in Nonlinear Elasticity and Piezoelectricity. *Applied Mechanics in the Americas*, Vol. I, pp. 477–482, American Academy of Mechanics and Asociacion Argentina de Mecanica Computacional, 1995.

Yu, Y.Y. (1995b) On the Ordinary, Generalized, and Pseudo-Variational Equations of Motion in Nonlinear Elasticity, Piezoelectricity, and Classical Plate Theories. *Journal of Applied Mechanics*, Vol. 62, pp. 471–478.

Yu, Y.Y. (1995c) Some Recent Advances in Linear and Nonlinear Dynamical Modeling of Elastic and Piezoelectric Plates. *Journal of Intelligent Material Systems and Structures*, Vol. 6, pp. 237–254.

Index